Chimpanzee and Red Colobus

Chimpanzee and Red Colobus

The Ecology of Predator and Prey

CRAIG B. STANFORD

With a Foreword by
Richard Wrangham

Harvard University Press
CAMBRIDGE, MASSACHUSETTS, and LONDON, ENGLAND 1998

Printed in the United States of America

Text design and typography by
Scott-Martin Kosofsky at The Philidor Company, Boston.
Set in Philidor Fairfield and Metro types.

Library of Congress Cataloging-in-Publication Data
Stanford, Craig B. (Craig Britton), 1956–
Chimpanzee and red colobus : the ecology of predator and prey / Craig B. Stanford
p. cm.
Includes bibliographical references and index.
ISBN 0-674-11667-4 (cloth : alk. paper)
1. Chimpanzees—Ecology. 2. Chimpanzees—Behavior. 3. Red colobus monkey—Ecology. 4. Red colobus monkey—Behavior. 5. Predation (Biology) I. Title.
QL737.P96S725 1998
599.885'153—DC21 98-2580

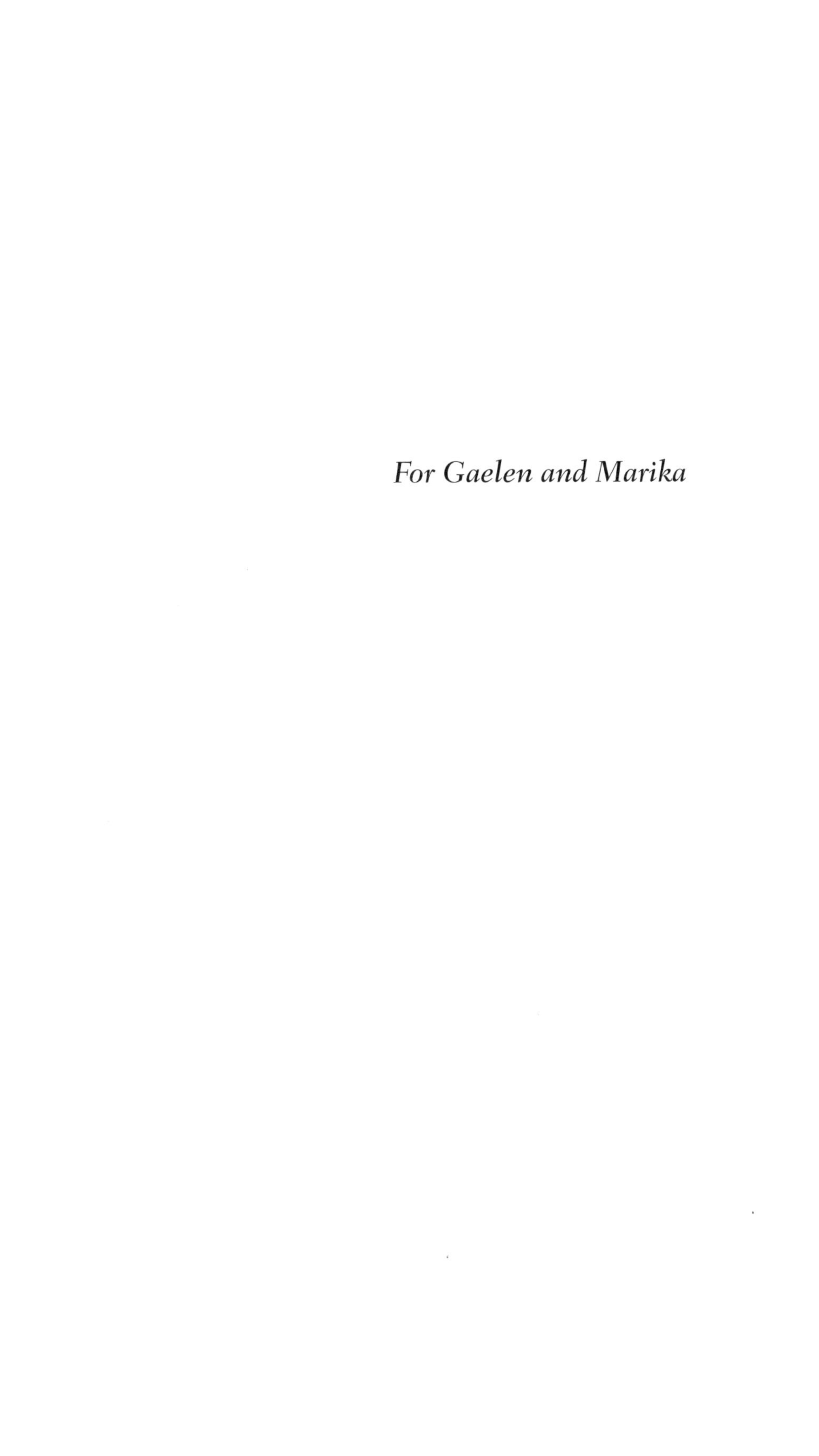

For Gaelen and Marika

Contents

Foreword

CHIMPANZEE THE HUNTER! Wouldn't Darwin have been delighted? Although he traced human ancestry to a fruit-eating African ape, a century passed between the publication of the *Origin of Species* (in 1859) and the first close-up field study of chimpanzees, so Darwin never knew how humanlike their behavior was. Then, wherever people watched, chimpanzees proved to hunt mammals, dramatically echoing the supposed lifestyle of prehistoric humans.

A hunting living ape had not even been imagined. Indeed, according to the dominant view of the 1960s, hunting was what made us human. It "had made men and women out of our apelike ancestors, instilled a taste for violence in them, estranged them from the animal kingdom, and excluded them from the order of nature" (Matt Cartmill, *A View to a Death in the Morning: Hunting and Nature through History* [Cambridge: Harvard University Press, 1993], p. 14). Thus the discovery of chimpanzee carnivory broke one more idea of human uniqueness. It suggested instead an important new dimension of ape-human continuity, one with special importance for ecology and behavior.

Though the basic facts about chimpanzee predation are now widely available—even to viewers of television documentaries—the larger meanings of chimpanzee predation have barely been explored. Craig Stanford came to this research fresh from a study of capped langurs (a species of monkey from Bangladesh), close relatives of the red colobus monkeys that make up the main prey of chimpanzees. He therefore began by looking at hunting from the prey's perspective, which no one had previously done. The result is a major advance in our understanding of the ecological role of chimpanzees as predators, and a magnificent platform for speculating about evolutionary significance. So this book addresses three big issues: the significance of predation for the prey, for the predator, and for human evolution.

Patterns emerge slowly in the study of chimpanzee predation because kills can be infrequent. Stanford solved this problem by working in Gombe National Park, where he was able to combine his own observations with a huge database on chimpanzee kills compiled by Jane Goodall's team over three decades and more. The result is a clear picture.

Chimpanzee predation is mostly a male activity, it covers a wide range of prey species, and it can happen in binges, when a community averages more than one monkey kill per day for weeks at a time. Styles of hunting and choices of prey vary between populations, but everywhere chimpanzees show tremendous excitement at both killing and eating other mammals. We can now be rather confident in our answers to questions about total rates of killing and meat eating, the impact of chimpanzees on red colobus groups, and individual and temporal variation in kill rates. The kill is a highly social affair, often leading to an initially chaotic division of the prey followed by a period when those who possess the meat can barter bits of it to influence beggars.

But intriguing puzzles remain. Why chimpanzees kill, and what their killing means for human evolution, would be easier questions to answer if we understood the nutritional significance of meat for this frugivorous, insect-eating ape. Is it calorie gain, or a specific nutrient, or a mixture of various nutrients, such as protein, fat, salt, or vitamin B_{12}? And what explains seasonal variation in kill rate? Are kills more common when plant foods are rare or when they are more abundant, and either way, why? Why are monkeys hunted by one species of chimpanzee but not, apparently, by the other (the bonobo)? Why do populations vary in the species that they eat? Why is there sometimes surplus killing, whereas at other times every last drop of blood is licked off the leaves? And do males really kill in the hope of exchanging meat for sex?

Chimpanzee and Red Colobus shows the excitement of the current, golden age of biological anthropology, reveling in the discovery of exciting new patterns and in their links to the big questions of where we come from and what we are. It teases us by showing the strangely conflicting pieces of evidence that speak to such questions, and by presenting clear ideas about them.

And like any study of apes, it reminds us that time is running out. It is already clear that chimpanzees have cultures of predation that vary from site to site, but those who would present data comparable to Stanford's must get themselves quickly into the field. Stanford's look into the strange carnivory of our cousins evokes a world of research opportunities that is disappearing faster than data are emerging. Man the Hunter! Darwin would not have been so happy about that. But he would have loved this book.

—Richard Wrangham

Preface

In this book I examine the ecological dance between two primate species, one of which is a predator on the other. This relationship creates an opportunity to explore the evolutionary role of predation in both the ecology and the social system of predator and prey species.

I took advantage of this situation by observing red colobus monkey groups in a forest in East Africa, learning the individual identities of the animals and monitoring the changes in the colobus population due to predation over the five years from 1991 to 1995. At the same time, I studied a population of chimpanzees habituated to researchers, which allowed me to follow both the hunters and their prey and to study the hunt itself from the viewpoint of both colobus and chimpanzee. This approach proved invaluable in testing hypotheses about how colobus monkeys respond behaviorally to the risk of attack.

My central goal was to understand both the proximate effects of predation on red colobus monkeys and the evolutionary influences on behavior and on population biology that may have arisen over a long period of time out of this intense hunting pressure. What began as an investigation primarily of red colobus behavioral ecology grew into one of the few studies of primate predator-prey ecology that has been conducted. For in the course of trying to understand the predator-prey dynamic, I began also to understand aspects of chimpanzee hunting behavior (and how hunting fits into chimpanzee society) that were unexpected and fascinating. My observational data on 119 encounters between red colobus and chimpanzees and on the response of the colobus to being hunted form the basis of this book, and are also the groundwork for a series of scientific papers published in biological and anthropological journals. In addition, I have made use of the huge database collected by the Gombe research assistants during periods when I was not in the field—nearly 300 additional hunts. This database is larger than exists from any other primate study site, and it has been invaluable in testing hypotheses about hunting patterns.

In addition to the data, I have included anecdotes and personal reactions to the research, because a full account of chimpanzee hunting and the courageous (no less anthropomorphic word appropriately captures their behavior) response of colobus monkeys is a compelling story. I have tried to keep the tables and graphs reasonable in number and unambiguous in content.

After Chapter 1 lays the theoretical groundwork, Chapter 2 describes Gombe as a field site—its animal inhabitants and the logistical problems of studying animals there. Chapter 3 presents the chimpanzees as individuals, and Chapter 4 summarizes what is known about chimpanzees as predators. Chapter 5 treats red colobus in their role as prey for chimpanzees. Chapters 6 and 7 discuss hypotheses about behavioral responses that the colobus might adopt to avoid predation, and look at what tactics are actually utilized. Chapter 8 examines the colobus population in combination with chimpanzee hunting patterns over the same period, and makes a preliminary estimate of the impact of predation on colobus population status and group size. It also presents a predator-prey computer simulation to predict future patterns of predation, and colobus population parameters using a number of possible variables. Chapter 9 asks the question "Why do chimpanzees hunt?" and attempts to reconcile the nutritional, social, political, and even reproductive incentives that appear to underlie the hunting behavior that I have observed. Chapter 10 explores a conceptual model of primate social systems based on the often-conflicting priorities of predator avoidance and food acquisition. Chapter 11 concludes with a summary and offers some implications for the future of this and related research on primate ecology.

Those looking for an exhaustive natural history of either the red colobus or the chimpanzee will be disappointed. That is not what this book is about. I have sacrificed information about many aspects of the life of each species in order to obtain information on the relationship between them. More space is devoted to colobus simply because much less is known and published about them than about chimpanzees.

Indeed, as I began this project there had been only one 9-month study of red colobus monkeys at Gombe, whereas Gombe chimpanzees had been watched for hundreds of thousands of hours by more than 50 researchers. In the 1990s this study, plus research by two doctoral students who also undertook long-term studies of colobus social behavior and ecology, have markedly increased what we know about red colobus. Nevertheless, some aspects of their behavior, such as feeding ecology, are given short shrift here, and I justify this unapologetically by resorting to a familiar excuse: you can do only so much with the time and observers available, so priorities are set. This book also deals only briefly with the implications of chimpanzee hunting behavior for understanding the behavior and ecology of our earliest hominid ancestors; I have presented my ideas and data on that topic in other publications.

Throughout this book I have tried to convey a sense of what this research entailed, and how the questions and hypotheses were formulated. The hunt itself is a behavior sequence as heartstopping to observe as anything in the field of animal behavior. Excitement is high not only for me, but also for the chimpanzees. Of the 60-odd hunts I have observed, in only a few could I see every bit of the action from start to finish. I have used detailed field notes from some entire hunts and from portions of others to explain and illustrate tactics of hunting and of colobus antipredator defense.

I need to say that I am interested in both the immediate and the evolutionary causes and in the implications of this predator-prey system. Since complex sociality is the most fundamental behavioral adaptation of the higher primates, it is likely that many aspects of the society of the red colobus are influenced by chimpanzee predation. Given a long history of being hunted by chimpanzees, natural selection has favored some behavioral strategies over others in coping with the risk of predation. The behavior of colobus upon being attacked (such as alarm calling or clustering together for effective defense) may involve proximate responses, but the tendency to carry out these actions is almost certainly evolved. Some readers may wonder how I can talk about the social system of the colobus or the chimpanzee as though it were an entity unto itself instead of the sum of the behaviors of many individuals. Although natural selection operates at the level of the individual (rather than at some higher level such as the population or species), when social behavior is involved it is often impossible to disentangle the actions of one animal from their impact on the other animals in the group. In Darwinian terms, the behavioral decisions made by the colobus, and by the chimpanzees that hunt them, are based on costs and benefits that influence reproductive success.

Two premises underlie all of my discussions. One is that there is probably no "unified field theory" that explains the workings of primate social systems, because neither food availability nor predation is likely to be the sole agent of natural selection in the wild. Instead, tropical forest ecosystems have a complex web of community relationships that depend on the abundance and body size of predators, their body size relative to primate prey, and the relative limitations of food and of predator pressure in those ecosystems. Only by understanding that web can we draw an accurate portrait of primate behavioral ecology and demography.

The second premise is that in nearly all cases we know very little about how wild primate populations are limited, because the energy requirements of the animals compared to the energy availability are almost never known, so we use indirect measures of this important information. To some extent, then, the field of primate feeding ecology rests on some rarely challenged but nevertheless untested assumptions. In spite of these caveats, it is possible to formulate some explanatory rules for the relative impact on primate social systems of food availability versus predation.

I make two general arguments in this book. First, I maintain that predation is a primary evolutionary influence on primate populations, but because it is much harder to document than food competition, it has rarely been studied. Second, I contend that the level of predation on red colobus monkeys at Gombe is not higher than on many other wild primate populations; it simply is more observable because both predator and prey are accustomed to human observers and thus can be watched continuously. This is almost never the case for any other predator and its prey. I state this at the outset, because the reader may ask whether the predation pressure on Gombe red colobus is an anomaly of that place or these species. As we shall see, that is not likely to be the case.

The research is presented here as though it proceeded from a well-organized plan. Nothing could be farther from the truth. In any research project, question A leads to questions B, C, and D, and addressing each question involves revising or rejecting one's original assumptions and therefore one's initial hypotheses—even more so in this case, because of two factors. The research was initially focused on the colobus themselves, and on the chimpanzees only with regard to their impact on colobus behavior. Once it became clear that I was able to use my data to address questions about chimpanzee behavior—Why do they hunt at all? What is their role as predators?—the study changed and I spent more time watching both species. As I returned to Gombe for each successive field season, I brought ideas for new directions, and for projects-within-projects. The resulting work has been nonlinear, but I believe has benefited from this approach.

Primates are the most easily and enjoyably watched forest mammals. They are large, active in the daytime rather than at night, and highly social. These qualities, along with the obvious comparisons to human behavior, make for easy fascination. But the reader should remember that this book is more about the evolutionary ecology of predators and their prey than about either of these primates—they are simply excellent examples of an evolutionary process that is interesting, important, and not very well understood.

Chimpanzee and Red Colobus

[Nature] arms and equips an animal to find its place and living in the earth, and at the same time she arms and equips another animal to destroy it.

—Ralph Waldo Emerson, *Essays* (1876)

1. Primates as Predators and as Prey

A PREDATOR AND ITS PREY are engaged in a mortal struggle. The prey live in big social groups, and a large percentage of the infants produced each year are destined to become fodder for the predators. The prey are attacked by a variety of predators, but one species in particular takes the heaviest toll. The predators travel in social groups, and when they encounter a group of prey their efficient hunting abilities are likely to spell death for their victims. They are able to hunt alone, but most often they attack as a group, and elements of cooperation in their strategy may enhance the odds of success. The act of killing can be a gruesome affair of tearing flesh, screaming prey, and satisfied grunts from the predators as they feed on the victim. The predators are so good at this business of hunting that they exert a profound influence on the age and sex composition of the prey population. In fact, natural selection may have molded the social behavior of the prey species through a long history of coping with the constant threat of attack.

The above passage could be describing the predator-prey relationship between lions and their wildebeest prey on the African plains, or between wolves and moose in palearctic boreal forests. In this case, however, both the predators and their prey are primates. The predator is the chimpanzee (*Pan troglodytes*) and its prey is the red colobus monkey (*Colobus badius*). These two species inhabit the same forest tracts across a wide expanse of equatorial Africa, and in all places where both occur and where both have been studied intensively, chimpanzees are known to prey on red colobus. The relationship between the two species is not analogous to the other predator-prey systems that are more familiar, because chimpanzees are not strongly dependent on the colobus population for food. Extinction of moose populations might spell trouble for wolves, and there are many records of declines of predator populations following prey population crashes.

We have little reason to think that red colobus are essential to the chimpanzee diet. Unlike wholly carnivorous hunters, chimpanzees are largely herbivorous, eating mostly ripe fruit. Only a small percentage of the chimpanzee diet is animals that they catch, though they clearly relish meat, bones, and blood. Red colobus, on the other hand, are strongly affected by the presence of chimpanzees. Many aspects of colobus behavior and ecology, from grouping patterns to life history, are subject to natural selection as a result of being attacked so often by chimpanzees. An omnivorous ape is interacting with its favorite source of meat. Thus, the predator-prey relationship provides a primate biologist with two research projects simultaneously: to discover how and why chimpanzees hunt for meat, and to understand the impact of this hunting on the behavior, ecology, and population biology of their monkey prey.

The very first successful hunt that I ever saw exceeded all my expectations for fascination and excitement. It was an utterly chaotic, 20-minute blood sport of apes intent on their quarry; branches broke, the chimpanzees screamed in excitement, and the colobus gave endless high-pitched alarm calls. What I remember most vividly is my own heart thumping as I tried futilely to follow the action overhead, and feeling the hair on my neck stand on end as an unseen capture was made to a chorus of electrifying screams from the hunters. The chimpanzees pursued time and again, lunging and then retreating in the face of courageous counterattacks by male colobus. During that first hunt I did not witness the kill itself, only its aftermath: a cluster of black apes on a high tree limb making animated attempts to get close to the captor, who held the limp, headless body of a baby monkey. The chimpanzees' hands extended in supplication near the captor's lips while he sat munching contentedly on the carcass. The sounds of bones crunching, lips smacking, and satisfied grunts from the chimpanzees were clear even from 20 meters below. Later the hunting party descended to the ground and sat close by, begging the captor for scraps of flesh from the skull—all that remained of the monkey. Objective data collection had been my goal that day; but I found it impossible to sit among the chimpanzee hunters tearing up the carcass and politicking for scraps of meat and not see the ghosts of our earliest human ancestors as well.

Nearly lost on me in that first hunt was that the colobus group from which the victim had been captured was still scattered through the treetops above, continuing to alarm call and watch their hunters feast. This was a small conquest to the chimpanzees, perhaps a kilogram of meat and bone divided among more than a dozen 45-kilogram diners. To the colobus it was the loss of one member of their social group, and to the mother of that monkey it was the loss of two years or more of time and energy in her prospective genetic perpetuation. Those two years were a substantial portion of her own reproductive life span, gone in a few frantic moments.

As with many other species linked in their relationship as predator and prey, this web was probably an ancient one. From the start of the interaction it was obvious that the mutual relationship was more complex and dynamic than I had expected, and that there were few easy answers to the questions I had brought to Tanzania. Those questions concerned not only the relationship between chimpanzees and red colobus monkeys in the forests of Africa where they co-occur: it is extremely likely that they apply equally well to the predator-prey ecology of many primates species elsewhere.

Studying Predation

At the outset of this study, I posed the following set of questions about the predator-prey ecology of chimpanzees and red colobus monkeys:

1. *What is the ecological role of chimpanzees as predators? To what extent does their hunting influence the size of the red colobus population, the size of individual red colobus groups, and the age and sex composition of those groups?*

The focus of most accounts of chimpanzee predator-prey ecology—all of which had been written by chimpanzee researchers—was the behavior of the predators. In the 1960s, research in Gombe National Park, Tanzania, by Jane Goodall and Geza Teleki, one of her early students, detailed the hunting behavior of chimpanzees for the first time. Throughout the 30 years of data collection at Gombe that had preceded my own study, researchers and their assistants had routinely recorded hunts. Little was known about the ecology, behavior, or population biology of any of the prey species except for baboons. Although baboons had been the primary prey of chimpanzees in the early years of Gombe research, during the past two decades they only rarely found themselves in that role.

The predator-prey dynamic of chimpanzees and red colobus had never been thoroughly investigated. Some researchers had estimated the intensity of predation from small samples of observed hunts and extrapolations from colobus population estimates (Clutton-Brock 1975; Busse 1977; Wrangham and Bergmann-Riss 1990). But little long-term or systematically gathered information existed on the precise effect of hunting on the red colobus population. It was known, for instance, that at Gombe most of the colobus eaten by chimpanzees were juveniles and infants. The effect of this pattern on the current and future age structure of the colobus population could not be ascertained without detailed information on the colobus. Although it was known that in some years hunting was more frequent than in others, the reasons (and the potential effect on the colobus population) were unclear. The size of each group probably would also be influenced by the immediate effects of intense hunting. Vigorous predation might also favor group sizes that were advantageous for avoiding attacks. If groups were of varying sizes, were smaller groups attacked more often than larger groups, or vice versa?

Chimpanzees use some portions of their territory more intensively than others. Was hunting therefore more intense in those areas, and if so, what was the impact on the red colobus groups there relative to areas less used by chimpanzees? Was hunting frequent enough in some years to severely reduce population size, or even to cause the extinction of some groups? None of these questions had been addressed for these species.

2. *What influences the hunting behavior of the chimpanzees?*

As the study of chimpanzee hunting behavior progressed through the 1970s and 1980s, evidence emerged of a behavioral diversity that paralleled the cultural diversity of the most technologically simple human societies. Different styles and features of hunting were reported from the various study sites: hunting in certain seasons in some forests but not others; a preference for adult prey at some sites but infants at others; cooperation during the hunt versus selfish noncooperation. Whether these differences were cultural or produced by local ecology, or even genetically influenced, was unknown. The intersite diversity of hunting patterns made unraveling the causes of hunting behavior at each site an important task. Every wild chimpanzee population that has been studied eats meat avidly, whereas hunting is rare or absent among the other three species of great apes: the bonobo, the gorilla, and the orangutan. Furthermore, chimpanzees at all four of the longest-running research sites in equatorial Africa—Gombe and Mahale in western Tanzania, Kibale in southwestern Uganda, and Taï in western Ivory Coast—took red colobus as their favored prey.

The two categories of potential influence on chimpanzee hunting behavior are social and ecological. Social factors arise from particular aspects of the local chimpanzee population and community. If hunting is more likely to be successful when the hunting party is larger, then factors that augment party size should increase the impact of hunting and predation on the colobus population. Since it is primarily males who hunt, factors that bring males together should increase hunting rates. Variation among individuals may also affect hunting patterns. For instance, if one particularly skilled or avid hunter lives in a given community, hunting may be more frequent than it would otherwise be; and if that hunter prefers infant colobus to adults, then the pattern of hunting by the community may change as a result. Broader social dynamics may also be at work, such as alliances between two or more hunters that increase hunting success and therefore trigger more frequent hunting.

Ecological factors, those in the physical environment, may influence hunting directly (if a lack of food or certain nutrients in some seasons is the impetus for acquiring meat) or may change social habits of the chimpanzees in ways that promote hunting (as when fruit is abundant and larger foraging parties form, which are then more likely to hunt). In addition, the habits of the prey themselves may exert a strong influence, as when a birth season for prey species produces a supply of babies at the hunting peak. Do the prey use a particular part of the forest or particular tree species at times that coincide with hunting peaks? With a large enough sample of encounters between chimpanzees and their prey, it should be possible to tease apart many of these potential influences on hunting behavior.

3. *What do red colobus do to reduce the risk of being attacked by chimpanzees?*

Since nearly all field studies of chimpanzee hunting behavior have been conducted by observers who studied only the chimpanzees, there are few detailed records of what colobus do in the minutes before they are approached by their predators. Even after an attack began, little was known about the behavioral response of the colobus—other than the obvious aggressive counterattacks launched by males. One would expect that red colobus would evolve responses particularly suited to their major predator's hunting technique.

I expected that members of a colobus group, in order to avoid potentially deadly encounters, would:

a. Avoid the same areas of the forest or the same plant-food species that are sought by chimpanzees.
b. Prefer to live in small groups to avoid being detected by chimpanzees or, alternatively, in large groups to reduce the odds for each of being the colobus who is captured. In either case, individual animals could make choices about the group size they preferred and thereby influence the grouping pattern observed at Gombe.
c. Be vigilant for predators, and try to maintain proximity to other group members to provide some protection from predation.
d. Use alarm calls that have been shaped by natural selection to give antipredator warnings. Studying the contexts in which alarm calls were given might reveal important details about colobus antipredator strategy.
e. Exploit the predator-detection skills of other vigilant monkey species in the same forest. Would it not be beneficial for the different species to associate and travel together in order to receive the benefit of additional vigilance? Or would doing so increase the overall group size so much that these multispecific groups could be located more easily by their predators? Do these species recognize the meaning of each other's alarm calls? Traveling together might be better for some species than for others, depending on their risk of predation, and this might be reflected in their pattern of socializing with other monkey species. Each decision by the colobus might

entail balancing behavioral benefits in one area with costs in another.

When red colobus fail to avoid a predator encounter, they must deal with a predatory attack. Male colobus counterattack as aggressively as songbirds mob a marauding cat. When male colobus counterattack together and leap onto chimpanzees in an effort to drive them away, are they acting cooperatively or simply making uncoordinated individual decisions? Social groups with multiple males are rare among colobine monkeys, the taxonomic group to which red colobus belong. If joint defense against attacks by chimpanzees is a behavioral adaptation, then chimpanzee predation might be responsible for the multimale society red colobus monkeys inhabit.

4. *Why do chimpanzees hunt?*

None of the other great apes hunt with the regularity chimpanzees do, and there is no evidence that they suffer nutritionally from a corresponding lack of animal protein. But chimpanzees spend much time and expend much energy in the pursuit of often-tiny quantities of meat in the form of small mammals. Chimpanzees could forage for plant foods in the forest at much lower cost of time and energy (and much less risk of injury) and these foods would provide many more calories per hour than meat does. The key question of why chimpanzees hunt had never been examined using actual field data.

As I accumulated observations of encounters between colobus and chimpanzees, I was able to investigate the several theories offered in the past to explain hunting. In Chapter 9 I present the hypotheses that might elucidate hunting patterns in chimpanzees at Gombe and elsewhere, and attempt to test them using available data from my own sample of observed hunts. A basic assumption in this discussion is that behavior is influenced at various levels simultaneously, and that evolved patterns of behavior, such as the tendency for all chimpanzee populations across Africa to hunt whereas some of their closest relatives do not, constitute one such level.

The question of why chimpanzees hunt thus addresses not only the immediate hunting decisions that are made each day, but also the role that hunting plays in chimpanzee society. If meat is desired for social as well as nutritional reasons, if it is used as a political tool or bartered prize, then in studying hunting by chimpanzees we are also looking deeply into the roots of meat-eating in the ape/human phylogeny.

5. *Has the risk of predation influenced the evolution of primate social systems?*

If the avoidance of predation is one of the basic requisites of life as a primate, then it should be considered a potential agent of natural selection—one that influences the behavior of individuals and the shape of primate social systems. It should also be possible to examine a range of behaviors that can be performed socially or individually, and test what antipredator benefits may accrue to animals that are social. For instance, does foraging in larger groups protect a red colobus from the risk of predation? If so, how? Does joining with other male colobus to defend the group during a chimpanzee attack increase the success of the defense?

These questions can be answered with the data presented in this study, though some potentially confounding factors must be isolated before we can assume that predation is the factor responsible for the behaviors observed. For example, the multimale grouping pattern of red colobus occurs in a variety of mammals. In some species, related males bond together to control females or to protect territories from males outside the group. Affirming that multiple-male groups exist for antipredator purposes requires unambiguous evidence that other factors are not likely to be be responsible for the grouping pattern.

THE EFFECTS OF PREDATION on a wild primate population have not been studied systematically, for reasons that are largely pragmatic. Unlike many common and easily observed behaviors of monkeys and apes, being eaten by a predator is a rare event. It occurs no more than once in the lifetime of any individual animal, and only a very few times in the life of a group over a field study of average length. Even if predator-caused deaths were a major cause of mortality in a wild primate population, the long-lived nature of the animals could mean that in the course of the typical 18-month doctoral project the observer might see only one predation event.

Moreover, in order to study predation one has to be able to watch not only the prey but also the predators, a feat that in most cases is impossible under naturalistic conditions of observation. Most animals who eat primates fall into one of two classes: the big cats and the larger birds of prey. The former are (with a few exceptions) solitary and largely nocturnal. Aerial predators are also usually solitary hunters, and quite shy of humans. None of these animals can be habituated easily to the presence of a human observer in order to watch them hunt the primates of one's study group. For precisely this reason studying the relationship between chimpanzees and red colobus is attractive to someone interested in predator-prey interactions. Chimpanzees are diurnal, highly social, and in some forests have been habituated to human presence, so that they can be followed on their daily travels and watched to see what they do and which animals they hunt.

An example of the difficulty in choosing predators to study comes from the Kibale Forest in Uganda, where Thomas Struhsaker and a team of researchers and students tracked red colobus groups for nearly two decades. In thousands of hours in the field, *no* confirmed predation by chimpanzees on colobus was seen, and only two predation events of any kind were observed. In later years, analysis of prey remains found under the nest of the powerful crowned eagles (*Stephanoetus coronatus*) indicated that raptors did prey intensively on the primate community, but no predation by chimpanzees was seen (Struhsaker and Leakey 1990).

It is likely that chimpanzees were eating red colobus throughout the study, but attacks were not observed because the predators were not habituated. They could not be followed and might have been deterred from attacking when humans were near. The Kibale chimpanzee population was later habituated, and hunts of red colobus began to be seen regularly (Wrangham, personal communication). The lack of data on predation was the result of focusing on the monkeys; studying only the colobus would not suggest frequent mortality due to predation unless one had information about the predators.

Indeed, Michael Ghiglieri (1984), studying a nearby chimpanzee community in an adjacent section of the same Kibale Forest, reported that chimpanzees there did not eat more than a trivial amount of meat. These chimpanzees were not well habituated, and only after they became more approachable by humans did anyone observe hunting behavior. Clearly, in both of these cases hypotheses about the behavior and ecology of both the red colobus and the chimpanzees might have been framed and tested differently had the researchers been aware of the extent of predation.

Some evidence exists that the mere presence of a researcher discourages a predator from engaging in its normal hunting patterns. In Amboseli National Park in Kenya, leopards are major predators on vervet monkeys, *Cercopithecus aethiops*. A study of vervet behavior there during the 1980s found strong circumstantial evidence that when human researchers were in the field following the monkeys, the local leopard or leopards changed their hunting patterns in order to avoid the humans, causing predation to drop in frequency (Isbell and Young 1993). Lynne Isbell (1990) estimated that nearly half of her vervet monkey population was eaten by one or more leopards in one year of her study, although not a single predation was witnessed.

It is plain to anyone who has walked through a forest that many of the animals present, including most predators, avoid contact with people. Feline predators like leopards are also nocturnal, whereas field researchers tend to be diurnal. Thus it is possible that, in many field studies, predator pressure on primates is relaxed during the period of research because of human presence.

In a field study I conducted on the capped langur monkey in a forest in Bangladesh, one aged female began to travel on the ground rather than through the treetops with the rest of the group. Over the course of my year and a half with the group, she spent more and more time on the ground, perhaps (in hindsight) because she became accustomed to the effect of my presence in keeping potential predators away. A month before my study concluded, she was ambushed and killed on the ground by a trio of jackals while I was 50 meters away looking for other members of the group. She may have become a prey item because of her habituation to an absence of predators, and this behavioral shift betrayed her one day when I was not near enough to keep them from approaching her (Stanford 1989, 1991).

Why Be Social?

As first light breaks over a forested valley in Tanzania, members of a red colobus group are awake, sitting in clusters of threes and fours and munching on a leafy breakfast. In the same valley a combined group of frenetic red-tailed and larger blue monkeys are combing a vine tangle high in a tree, looking for tiny fruits and insects while scanning the sky overhead and the ground beneath for possible danger. At the same time a few hundred meters away, several baboons are climbing out of their night trees and marching down to the beach to scour the sand for bits of food, arguing, socializing, and copulating as they go. And at the same moment, high in the same valley, a party of chimpanzees are on the move, their morning having begun with a charging display by the alpha male followed by much appeasing and reconciling.

Primates are, above all else, social creatures. This is in contrast to the majority of other mammals, who tend to be solitary and avoid one another except for relatively brief periods of contact for mating. Three primary factors are believed to promote group living: access to mates, proximity to and defense of food sources, and protection against predators. Since most primates live in groups, we might assume that some combination of these three must outweigh the disadvantages, and in fact this is the premise that most primate ecologists employ. The assertion is based on our current understanding of the fundamentals of social systems and grouping patterns of monkeys and apes.

Social systems are adaptive to the individual. Just as many other aspects of an animal have evolved under the pressure of natural selection in a specific environment, the animal's inclination to live in a certain kind or size of group is likely to have an evolved basis. The form and behavior of an animal is a mosaic of adaptation to both past and present circumstances in its unique evolutionary history. In many primate species the social system appears quite rigid even when the environment differs dramatically from one region to another in the species' geographic range. The basic strategies of survival and reproduction, including what kind of group to live in, therefore have evolutionary bases.

There is a paradox, however, in this sociality. In the 1930s, the British zoologist Solly Zuckerman watched hamadryas baboons who lived in the London Zoo. Like most baboons, they spent enormous amounts of time fighting, courting, and copulating. Zuckerman (1932) concluded that the most basic need of a primate is for a ready source of potential mates, and that access to mates is best provided by living in social groups. Indeed, since the majority of higher primates live in groups containing both males and females, mate availability would not seem to be a problem for either sex. But a male who seeks females in a large social group must deal with the down side of living in a group: other males. A male who theoretically has the opportunity to mate with all of the females in the group may find himself excluded by his virtue of his low dominance status, his inability to gain access to the females at key times (such as when they are ovulating), or his failure to be chosen as a mate by the females themselves.

Predation also produces conflicting priorities with regard to group living. The benefits of living in a group seem obvious: more eyes to watch for the eagle that swoops in from above or the leopard that creeps up from below. In some species, some or all group members are able to repel a potential predator physically with a mobbing counterattack, as when songbirds drive off an owl. For all but the smallest-bodied primates, large primate groups should be more effective at this antipredator defense than small groups. But at what cost? Might not larger groups be more easily detected, and so have their members' individual safety compromised by virtue of their greater overall visibility? Is there a point at which the benefit of having the protection of more and more group members becomes a liability? Is the danger of being eaten strong enough to influence the size and the type of group in which the animal lives? And, perhaps most important, there are likely to be conflicts for primates as they try to balance the basic priorities of sex, food, and predator protection.

The relative importance of predator avoidance as the reason for being social is the central theme of this book. In a review of antipredator behavior in birds and mammals, Lima and Dill (1990) point out that failure to mate, failure to gain access to food, and failure to avoid predators carry quite different potential costs. If an animal tries to mate one day and fails, the negative effect on its overall lifetime reproductive output is small; many mammals mate thousands of times for each conception. The same is true for food; eating poorly for a day or a week probably does not have a measurable negative impact on a long-lived animal's overall reproduction. But failing just once in a lifetime to avoid being caught and eaten by a predator puts an end to all future possibilities for offspring and of course to life itself. Both Geerat Vermeij (1982) and Robin Dunbar (1988) have suggested that just the *risk* of being eaten may be as important an evolutionary pressure as actually being eaten, since by avoiding predation those best able to protect themselves or escape can pass their genes on to their descendants.

Predation can therefore be inferred to be an important influence on behavior. The precise degree of importance will vary depending on factors such as the predator species present (are they large enough to eat primates?), the body size of the primate species (are they small enough to be at great risk?), and the type of habitat (forest or savanna). Predation has immediate influences on an animal's hourly activity pattern. The risk of predation may have also produced evolved behaviors that exist because those ancestors who behaved properly in the face of predators enjoyed greater reproductive success than their neighbors who were eaten.

Feeding Ecology and Predation

If the first reason for the failure to study predation is logistical, the second reason is theoretical; its importance as an evolutionary force affecting wild primate populations has often been considered minor relative to the continual pressures of finding and defending satisfactory food sources. A survey of researchers conducted in the 1980s was inconclusive with regard to how widespread the influence of predation is among species of the primate order. Dorothy Cheney and Richard Wrangham (1987) found no significant difference in predation rates on primate species living in multimale groups as opposed to one-male groups, nor between arboreal and terrestrial species. Nor was a significant difference corresponding with habitat or ecological found in predation rates across species. No trend in predation patterns that might support an adaptive basis for particular grouping patterns was apparent.

In sharp contrast are the effects of diet and feeding ecology on behavior. Since the 1960s, the importance of diet and food acquisition in influencing reproductive success has been recognized and widely studied by primatologists. Studies of diet and food tend to produce more testable hypotheses in the course of a field study than does research about predation, because data on feeding can be collected readily. Most primates spend 30–50 percent of their daylight hours eating, so the quantity of data that can be collected in even a short study, combined with observations of the animals' behavior in association with feeding, permit tests of hypotheses about the effect of diet on behavior.

Most primate field researchers who have looked for competition in primate feeding ecology have found it. Most researchers have agreed

TABLE 1.1. *Some studies that have addressed the limiting factors on wild primate populations.*

STUDY	SPECIES	SITE	LIMITATIONS	NATURE OF EVIDENCE
Hladik 1975	*Lepilemur mustelinus*	Madagascar	Food	Estimate of food available in dry months
Janson 1985	*Cebus apella*	Cocha Cashu, Peru	Food, but seasonal	Competition, plus estimate of energy intake vs. needs
Milton 1980, 1982	*Alouatta palliata*	Barro Colorado, Panama	Food	High mortality of immatures due to seasonal food shortage
Coelho et al. 1976	*Alouatta palliata, Ateles geoffroyi*	Tikal, Guatemala	Not food limited	Estimate of food available vs. food eaten
Cheney et al. 1988	*Cercopithecus aethiops*	Amboseli National Park, Kenya	Food and predation	Ecological crunch due to tree die-off
Waser and Case 1981	*Cercocebus albigena*, several others	Kibale Forest, Uganda	Food	Aggressive interactions between species at food sources
Stanford 1995a	*Colobus badius*	Gombe, Tanzania	Predation	Group size and predation intensity inversely related
Marsh 1986	*Colobus badius rufromitratus*	Tana River, Kenya	Food Seasonal food limitation	Ecological crunch due to tree die-off
Wrangham 1975	*Pan troglodytes*	Gombe, Tanzania	Aggressive interactions between species at food sources	Weight loss in dry season

that competition for food, both within and between groups, exists and is a potential evolutionary influence on the size and structure of groups and of social systems. This conclusion has been arrived at in a number of ways. First, studies have shown that as the size of a primate foraging group increases, the per-capita food intake may decrease, suggesting that sociality brings a cost of competition (Wrangham 1975, chimpanzees; van Schaik 1983 and van Schaik and van Hooff 1983, long-tailed macaques; Janson 1985, *Cebus*). Second, plant foods contain particular nutrients, and we observe that some plant foods are eaten more frequently than other foods that contain fewer nutrients or fewer digestible parts. For example, chemical analyses of leaves eaten by colobine monkeys in both African and Asian forests showed that the least fibrous and most easily digested leaves are eaten most often (Waterman and Choo 1981). Third, we see members of one group, and also group members of different species, fighting for food resources (Waser and Case 1981).

Exactly what do analyses of primate feeding competition measure? Studies purporting to show the evolutionary importance of food competition both between and within primate groups are based on the assumption that food for wild primate populations is limited. Animals must partition their use of the available food, and therefore individuals have evolved specialties for utilizing different foods or different parts of the same food source.

In the case of a single species having finite resources, intragroup competition ought to occur. The ability of each female of the species to obtain these resources should be a crucial component of her lifetime reproductive output. When the food supply decreases, reproductive performance should decline as a result. But is this the case?

Table 1.1 surveys the relatively few primate studies that have examined the limits on wild primate populations. These include a few strong demonstrations, such as the observation of widespread mortality in a population at the same time as a crash in food resources. On Barro Colorado Island in Panama, for example, a dramatic decline in plant productivity in the 1960s was accompanied by a die-off of many monkeys that relied on those plants (Foster 1982). It must be noted, however, that a crash in the same plant resources a decade later did not result in such a population decline, at least among the howlers on the island (Milton 1982). A decline of the red colobus of the Tana River

Reserve in Kenya may have been linked to a drop in the availability of forest food trees, although the effect of tree decline was not felt quickly and studies of the colobus were equivocal about the causes of both tree and monkey population decrease (Marsh 1986, discussed in Davies 1994). Vervet monkey populations in Amboseli National Park also suffered precipitous declines at the same time that their food resources were unavailable due to a die-off of many trees (Cheney et al. 1988). A second indication of the link between food and survival is the rapid expansion of a primate population when its food supply is increased, as occurs with some human-provisioned Japanese and rhesus macaque populations (Mori 1979).

Strong circumstantial evidence of food limitation also exists. Charles Janson of the State University of New York at Stony Brook estimated the energy requirements of Peruvian brown cebus monkeys (*Cebus apella*) living in a seasonal forest in Cocha Cashu, Peru. Using estimates of the availability of plant food, he inferred that during the lean season food was indeed a limiting factor for cebus monkeys (Janson 1985). Food intake was limited by aggressive competition and dominance relationships among group members. Janson's is one of the very few field studies in which energy intake has actually been estimated (in kilojoules per gram) to determine nutritional or energetic constraints on the monkeys. Janson found food limitation in the form of an ecological crunch during the driest months. Feeding competition within each cebus group was greater than competition between groups by a factor of ten.

The notion that food may be limited during ecological crunches but not necessarily throughout the year has been put forward for many animal populations. The ornithologist John Wiens noted that feeding competition within groups occurs even in the absence of food limitation. Measures of feeding competition therefore should not be used to infer that an animal population lives constantly at or near the limits of its energy availability. Wiens (1989) showed that periodic food declines may be the sole time when food competition really matters. Natural selection may operate on wild populations only during such crunches: one season of the year, one year in a decade, or one year in a century. Hladik (1975) also estimated resource availability for a small Madagascar prosimian, *Lepilemur mustelinus,* and inferred that the population was food limited in the leanest season.

At least one field study has suggested that food limitation does not occur. Anthony Coelho and his colleagues, working in Tikal National Park in Guatemala, estimated that food abundance was greater than the spider and howler populations there require by many orders of magnitude (Coelho et al. 1976). John Cant (1980) later disputed these findings, noting that the short duration of the field research may have meant the study period was not representative of longer-term patterns.

Beyond these few studies, there have been many assertions and inferences about food competition but little evidence. In most studies it is simply assumed that competition between individuals for food within a group shows that food must therefore be limited. The limitation may in fact be due to patchy distribution rather than to overall lack of availability (van Schaik 1989). The degree of food competition is usually surmised from patterns of feeding behavior, even though the level of resource use does not necessarily indicate anything about competition. Wiens notes that even when fighting among individuals is caused by a food source, the intensity of aggression does not necessarily indicate the nutritional importance of the resource to the individuals unless all confounding social reasons for within-group competition can be eliminated.

Likewise, even when observing divergent diets among two different primate species, we should not assume food limitation as the cause of interspecific niche separation unless it can be shown that such partitioning would not exist if resources were not limited. Ecological studies of birds have addressed these questions for years, and an ongoing debate exists over whether food is limiting and competition orders interspecific niches and intraspecific dynamics (Diamond 1975), or whether such competition is usually absent or at least not linked to supposed food limitations (Strong et al. 1984).

My intention is not to heap undue criticism on studies of primate feeding ecology, but to suggest that theories of the effect of food limitation on wild primates are often based on poorly tested assumptions. Field studies should devise increasingly clever methods of measuring resources and their limitation thresholds. At the same time, field studies examining the significance of predation may become more useful in explanatory models of primate social behavior as longer studies compile better information. Whether ecological communities are structured more by resource availability ("bottom-up" organization) or by predators ("top-down" organization) is strongly debated among tropical ecologists (Wilson 1987; Terborgh 1988; Wright et al. 1994). Research on predation may provide an answer.

Only a few reviews have been made of published evidence for the importance of predation. The Cheney and Wrangham article found little support for predation as a pervasive ecological force shaping primate social systems. Isbell (1994) concluded that arboreal primates are more vulnerable to predators when in exposed habitats, such as forest edges and above the forest canopy. She showed that predation rates, expressed as the percentages of mortality to the study population annually that are due to predation, are positively correlated with distance from human habitation and somewhat negatively correlated with group size.

Anderson (1986) examined published evidence of predation to test for positive correlations with other features of prey populations. She tested seven hypotheses about the effect of predation on wild primates, and her results failed to reject any of the seven. High predation levels were positively correlated with (1) the most sexually dimorphic species (by body mass), (2) species in which males tended to defend their conspecifics against predators, (3) group size, (4) low frequency of lone animals, (5) low rates of female transfer, (6) lesser tendency for group fission, and (7) many adult males per group. The data were taken from studies of 102 populations of 31 primate species, although many of the observations were anecdotal and mortality rates from predation were reported for only 11 populations. No attempt was made to separate variables that may themselves be correlated, such as the number of males per group and overall group size. Some of Anderson's hypotheses are likely to be artifacts of others. She has, however, put forward suggestive evidence that predation influences some of the characteristics of primate societies.

An additional review was published by Goodman et al. (1993), who compiled anecdotal records of predation on Malagasy prosimians. These researchers found that in spite of the absence of large raptors (though remains of an extinct species have been found) and large carnivores (the fossa is an extant medium-sized carnivore that eats lemurs), predation nevertheless affects many species.

Table 1.2 presents predation information for primate studies in

TABLE 1.2. *Published studies that have produced systematic data on predation on wild primate populations.*

SPECIES	SITE	ESTIMATED MEAN ANNUAL PREDATION RATE	SOURCE
Microcebus rufus	Beza Mahafaly Reserve, Madagascar	.25	Goodman et al. 1993
Propithecus diadema edwardsi	Ranomafana National Park, Madagascar	.08	Wright 1995
Callicebus moloch	Manu National Park, Peru	.04	Wright 1984
Aotus trivirgata	Manu National Park, Peru	.00	Wright 1984
Cercopithecus cephus	Makoku, Gabon	.10	Gautier-Hion et al. 1983
Cercopithecus aethiops	Amboseli National Park, Kenya	.30–.60	Cheney et al. 1988
Papio cynocephalus	Amboseli National Park, Kenya	.10 (minimum)	Altmann and Altmann 1970
Papio anubis	Gombe National Park, Tanzania	.065	Ransom 1981
Papio ursinus	Moremi, Botswana	.08	Busse 1980
Colobus badius	Gombe National Park, Tanzania	.04–.06	Busse 1977
Colobus badius	Gombe National Park, Tanzania	.08–.13	Clutton-Brock 1975
Colobus badius	Gombe National Park, Tanzania	.18	Stanford 1995a
Presbytis pileata	Madhupur National Park, Bangladesh	.075	Stanford 1989
Macaca sinica	Polonnaruwa, Sri Lanka	.01	Dittus 1975
Pan troglodytes	Mahale National Park, Tanzania	.06	Tsukahara 1993

which researchers obtained direct estimates of (1) the percentage of primate prey in the total diet of the predator, or (2) the percentage of the primate population that was eaten by the predator annually. This compilation omits many studies cited by Anderson (1986) that are based on anecdotes or impressions (such as a researcher's conclusion that "predation was moderate"). The table includes two recent investigations not cited in Cheney and Wrangham (1987) or Isbell (1994). Some of the studies report quite high rates of predation; predation mortality is more than 20 percent annually among some monkey populations (vervets, Cheney et al. 1988; red colobus, Stanford 1995a) and 25 percent annually in at least one prosimian (mouse lemurs, Goodman et al. 1993). In none of these cases is there evidence that the predation levels contribute to population decline or local extinction.

The evidence for an ecological or evolutionary role for predation in shaping primate societies is considered by many scholars to be highly circumstantial. For few species is there unambiguous evidence that predation is a primary influence on behavior or population biology. The reason may be that systematic data and documented predation rates are available for only a few wild populations. Nearly all predation data have been collected incidentally during studies focused on other aspects of the species' biology. In the few populations where predation events can be observed directly, both detailed descriptions of the pattern and an understanding of the long-term effects are possible.

The eastern shore of the lake, along which we coasted, was a bluff of red earth pudding'd with separate blocks of sandstone. Beyond this headland the coast dips, showing lines of shingle or golden-coloured sand, and on the shelving plain appear the little fishing villages. They are usually built at the mouths of the gaps and gullies, whose deep gorges winding through the background of hill-curtain become, after rains, the beds of mountain torrents.

—Richard Burton, *The Lake Regions of Central Africa* (1860)

2. An African Forest

Gombe National Park is hallowed ground in the history of animal behavior study. As the site of the longest-running study of any wild animal population, it is filled with places and animals familiar to those who have followed the work of Jane Goodall and the many students who came after her (see Appendix 1). Primatologically historic landmarks abound: the hillside where Goodall first saw chimpanzees eating meat; the rocky outcrop called the Peak, where Goodall watched her still-shy chimpanzees in the early 1960s; the places where chimpanzees were ambushed and killed by males of the northern chimpanzee community during the warfare of the 1970s. Every student of primate behavior over the past three decades has been influenced to some extent by the films and magazine articles that appeared in the early days of chimpanzee research at Gombe.

Before my first arrival, I knew of Gombe as a remote and rugged forest tract on the northeastern shore of Lake Tanganyika. Today Gombe is much less remote, surrounded on all sides by humanity. The rugged beauty is still its major feature to the newcomer, and there are few truly flat places once one leaves the lakeshore. The national park and its chimpanzees have an elevational range of 750 to 1,500 meters. The high-canopy rain forest that is the habitat of many primate populations across Africa does not extend to western Tanzania, which is seasonally arid. Gombe experiences a period from early June through October when little rain falls. The result is a mosaic of habitats ranging from grassland to rushing streams that emerge from high on the western slope of the Rift Valley escarpment, traverse the many steep-sided valleys, and empty into the lake. If Gombe were lifted from the map of Africa and set down among the mountains behind my home in southern California, it would not look terribly out of place. The weather is also warm, but rarely hot and never cold, creating a user-friendly climate for field research.

The forest is not, however, without major challenges to a follower of monkeys or chimpanzees. Most hillsides are carpeted with thorny vine thickets (*Smilax* sp.) that allow chimpanzees to slip through but entangle human legs and make tracking animals slow and difficult. Many trails are sheer and badly eroded, covered with loose paprika-colored gravel in the dry season and a layer of slippery clay mud during the winter rains. Chimpanzees stride up steep hills at about the pace that a human would walk on flat ground. Particularly in the case of hunting, many key events happen moments after the chimpanzees meet their prey and the hunt begins. I might be hiking, running, or crawling many meters behind. Even in the middle of a chimpanzee party, observing a hunt can be an exercise in frustration. The impenetrable thickets block all visibility, while the screams of chimpanzees and calls of colobus in the canopy above suggest the exciting events (and data) one is missing.

There are better days, however, when a hunt occurs in the canopy over an open streambed, allowing an observer to see the entire episode unfold from below. Occasionally a hunt takes places in the trees of a steep hillslope, so that the action can be observed from the ground farther up the hill with a magnificent eye-level view of the animals.

Gombe National Park is a tiny island in a human sea. Since the 1960s, the growing human population in the Kigoma region of the Lake Tanganyika shoreline has surrounded the park and effectively cut it off from other forest patches that remain in northwestern Tanzania. What remains is a region of forested mountains, 45 square kilometers in size (Figure 2.1).

Although it has been referred to as a savanna-woodland habitat, this description of Gombe is somewhat misleading. The savanna exists in the form of open grassy slopes dotted with *Brachystegia, Uapaca,* and other tree species on the higher slopes. This habitat type, known as miombo woodland, occurs widely in East Africa. The miombo is traversed by chimpanzees and also by red colobus when moving from one valley to the next. Frequent references to Gombe chimpanzees as living in savanna-woodland have been made by researchers attempting to depict Gombe as a habitat very similar to that in which our earliest ancestors evolved in the Pliocene (Goodall 1968) and also by

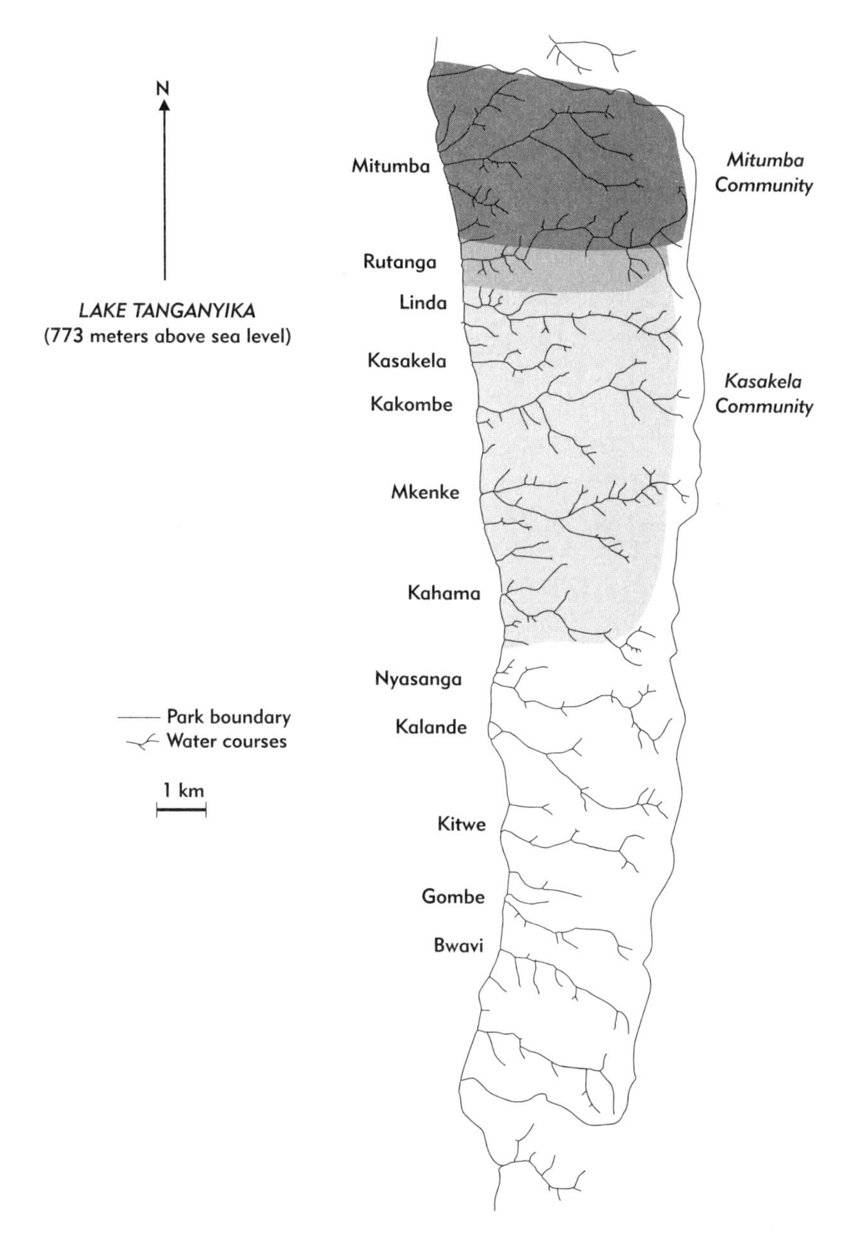

FIGURE 2.1. Gombe National Park, showing the territories of the Kasakela and Mitumba chimpanzee communities. Adapted from Goodall 1986.

researchers attempting to dichotomize chimpanzee habitats into wet and dry types (Boesch 1994b). The dichotomy is to some extent false, because the animals seldom use the open grassland areas. Below the miombo is a band of dense, often thicketed, semideciduous woodland on most of the valley slopes. Thickets surround an irregular canopy of trees penetrated by emergents such as *Albizzia* spp., *Newtonia buchanani,* and *Pterocarpus angolensis.* Streambeds, or *korongos*, cut each valley vertically. Most are dry ravines that carry mountain runoff after storms, but some streambeds in the study area—Mkenke, Kakombe, and Rutanga—have year-round rushing streams with potable water. Where the streambeds dissect the valley bottoms, the surrounding riverine forest is continuous canopy up to 30 meters in height. Chimpanzees spend much of their time in this habitat, mainly because of the abundance of fruiting trees—especially figs. Significantly, it is also the preferred habitat of red colobus, who spend long hours in the tall *Albizzia, Newtonia,* and streamside *Ficus* species.

Despite almost 40 years of research at Gombe, the forest itself has never received the attention given the animals. The site has been broadly described Goodall (1963, 1968, 1986), but no quantitative description of the forest was made until Anthony Collins and William McGrew (1988) conducted a vegetation analysis as part of a site comparison with the Mahale Mountains 160 kilometers to the south. Clutton-Brock and Gillett (1979) had also studied the vegetation in connection with Clutton-Brock's red colobus study in the late 1960s. The phenology of the forest has not been rigorously studied, even though it is commonly observed that many of the important chimpanzee food trees experience boom-and-bust years that seem to be related to rainfall patterns. These in turn appear to influence chimpanzee travel patterns (Wrangham 1975).

Gombe, like the rest of western Tanzania, has a highly seasonal climate, with two main rainy periods between November and May and a prolonged dry season the rest of the year. The forest's appearance changes dramatically between the wet and dry months. The pervasive dampness and lush foliage during the rains contrast with the scene during June to October, when the ground is covered with dry leaves, a warm breeze blows constantly, and visibility in the forest is excellent because of the lack of foliage. The behavior and ecology of both chimpanzees and red colobus also change from the wet to the dry months, as we shall see. Even within the small park a climatic gradient exists. The southernmost valleys of Gombe are drier with more extensive areas of miombo on the valley slopes, while the northernmost valleys are more lush. Some tree species that bear seasonal chimpanzee foods, such as *Myrianthus arboreus* (Moraceae), occur in the north of the park but not in the south. This species is also found mainly in the more luxuriant parts of Mahale National Park, suggesting that it is less tolerant to drier conditions. The climate gradient presumably exists because of the way in which points along the rift escarpment capture and trap moisture, and perhaps create more rainfall, in the northern end of Gombe.

For a variety of reasons, Gombe National Park is not rich in large-animal species. The forested area is small and surrounded on the north, south, and east by cultivated land, and on the west by Lake Tanganyika. About 240 bird species are present (Stanford 1994; Stanford and Msuya 1995), but many large mammals are not. Ungulates such as buffalo, waterbuck, and roan antelope occurred in the recent past but owing to poaching are now locally extinct (Goodall 1986). There are, however, numerous predators other than the chimpanzees themselves (Table 2.1); leopards are seen occasionally, and at least one chimpanzee death in recent years was attributed to a leopard kill. The leopard population must be very small given the size of the park, though large cat tracks and dung that I have seen on the rift escarpment suggest that the cats may hunt in both the forest and the farm areas outside the park. At least one crowned eagle nest exists within the study area, and large raptors are abundant, particularly in migration.

Eight nonhuman primate species occur (Appendix 2). Of these, two are seldom-seen nocturnal prosimians, five are monkeys, and the other is the chimpanzee. Chimpanzees, baboons, and red colobus have all been the subjects of long-term field studies. The black-and-white colobus *(Colobus guereza)*, which occurs farther south on the eastern lakeshore in the Mahale Mountains, does not occur at Gombe. The blue and red-tailed monkeys, still little studied, present fascinating questions about hybridization. Nearly every red-tailed monkey group seen at Gombe contains at least one hybrid of red-tailed and blue monkeys, with intermediate physical features and distinct vocalizations.

Within Gombe live at least three and perhaps four communities of chimpanzees, each occupying separate, overlapping territories (see

TABLE 2.1. *Known and suspected predators on primates in Gombe National Park.*

SPECIES	LIKELY MAIN PREY	CONFIRMED PREY	SOURCE
Leopard	All primates, other mammals	Baboons only	Hair in dung
Crowned eagle	All monkeys, other small mammals	Baboons and other unidentified monkeys	Wallis and Msuya, unpublished data
Martial eagle	All monkeys, other small mammals	None	—
Chimpanzee	Red colobus	Red colobus, other monkeys	This study and others
Olive baboon	Smaller monkeys	Red colobus	Goodall 1986
African python	All mammals	None	—

Figure 2.1). Two communities have been habituated and are being studied, the best known being called the Kasakela community. It comprises the chimpanzees habituated by Goodall beginning in the early 1960s, on whom a rich database of information has been collected. In 1996 it numbered approximately 44 individuals, including 9 adult males, 3 adolescent males, 11 adult females, plus immatures. The community has a territory that encompasses about 18 square kilometers. Overlapping this area to the north lives the more recently habituated Mitumba community, approximately 25 individuals occupying a smaller area of about 10 square kilometers. Some females in this community are not yet habituated to being followed, and new females are still being sighted. Following an outbreak of a respiratory ailment in early 1996, only 3 adult males remain. The territory of the Kasakela chimpanzees extends from Rutanga Valley in the north through Kahama Valley in the south; occasional forays take them even farther than these perimeters.

Within the range of the Kaskelas is an intensively used core area that is centered in Kakombe Valley. I chose this valley as the base for my study because it is where chimpanzees and colobus encounter each other most often. I followed chimpanzee parties out of Kakombe into the adjacent Mkenke Valley to the south and Kasakela and Linda valleys to the north. I also spent two months in Mitumba Valley, comparing the relationship of colobus and chimpanzees there with that in Kakombe.

In the same 18 square kilometers controlled by the Kasakela chimpanzees lives a large population of red colobus monkeys. I estimated that 500–550 of them inhabit this area, living in approximately 20 groups. Gombe red colobus live in multimale groups averaging 28 animals, though the size of group in different parts of the park varies by 100 percent. I wanted to study the behavior of at least one group in detail, as well as monitor several other groups for the potential effects of hunting on them. In the end, I studied two groups intensively and monitored three others. The five groups had overlapping home ranges in or near Kakombe Valley and so were all subjected to chimpanzee attacks, some of them repeatedly, during my time in the field. These groups were smaller than many other colobus groups at Gombe and occupied smaller home ranges than those estimated for other groups.

The Question of Human Influence

In the early days of research at Gombe, bananas were used to provision the then-unapproachable chimpanzees and hasten the process of habituation to human observers. Most of the observations were made in a cleared feeding station, where the chimpanzees gathered each day to wait for—and sometimes to fight over—the banana ration. Artificially heightened levels of intragroup aggression resulted (Goodall 1986), and in the late 1960s the provisioning was curtailed. Some scholars have questioned whether the behavior of chimpanzees at Gombe and also in Mahale National Park, which was also provisioned in the early years of research, may be different from naturalistic patterns of chimpanzee behavior. Power (1991) has even suggested that our view of chimpanzees living in societies driven by dominance among males, intense territoriality, and lethal intercommunity aggression may be an artifact of the provisioning in Gombe and Mahale. That claim is refuted by the behavior of other, nonprovisioned chimpanzee populations that exhibit the same pattern of intercommunity aggression seen at Gombe (Brewer 1978; Wrangham and Peterson 1996). Moreover, the unhabituated chimpanzees at Gombe are just as territorial and aggressive as the habituated chimpanzees; occasionally they are brazen enough to attack female chimpanzees in the presence of human observers.

This general topic raises a number of important issues not only for chimpanzee research, but for the study of wild primates and all other animals in human-influenced settings. Chimpanzee hunting patterns may have been colored by the provisioning of the chimpanzees during the 1960s (Goodall 1986). Baboons, drawn to the same banana supply, came in close contact with chimpanzees day after day, and the easy availability of baby baboons led to high levels of predation on them.

A few scholars have suggested that the hunting was aberrant behavior induced by human disturbance. This claim incited fierce debate (de Pelham and Burton 1976; Teleki 1977); at stake was whether hunting itself was natural chimpanzee behavior. Observation of chimpanzees hunting for a wide variety of mammalian species, including an occasional Gombe baboon, in all ecological contexts in the ensuing decades has silenced the argument. It is noteworthy that chimpanzees still encounter baboons frequently in the forest and at camp in Gombe, but rarely hunt them. Ransom (1972) and Goodall (1986) hypothesize that baboons gradually learned that chimpanzees were dangerous predators and consequently became more difficult prey. Chimpanzees may have preyed on Gombe baboons, at least occasionally, long before there was a human presence at Gombe. The reason that baboons are today hunted only a few times a year is ambiguous.

Arguably, no forest in the world today is free of human influence, which is of two potential types. People pursuing a traditional way of life, such as hunting and gathering, may affect the animals directly through hunting or indirectly through human changes to the landscape. There are also the more recent effects of people living in close proximity to primates, including humans who come to the animals' habitat specifically to learn more about them. Both of these factors, but especially the latter, present dilemmas for primatologists wishing to understand the naturalistic behavior of the animals in question.

The problem is more complex than eliminating direct interference. At Gombe the most important year-round plant-food species for chimpanzees is the oil palm (*Elaeis guineensis*). This tree is native to coastal West Africa but is cultivated widely in the East (Hartley 1977). It is used by some chimpanzee populations but not by others (McGrew 1992), and its highly nutritious fruits provide Gombe chimpanzees with nutrients that might not otherwise be available to them. It thereby exerts an influence on survival and reproduction that may be a much stronger, though less obvious, human influence than the effects of provisioning.

Such influences are by no means limited to chimpanzees. Mountain gorillas, for example, have been heavily subjected to poaching in recent decades, which has altered the natural social dynamics of groups that lost their silverback males (Fossey 1983). It is not clear whether one-male groups are the norm or occur mainly when human influence has removed many adult males. Mountain gorillas forage extensively for human crops and for introduced plant species surrounding their forest habitats (Stanford, personal observation). Langurs and other monkeys in India have been studied primarily in environments that have undergone hundreds or even thousands of years of human alteration (Hrdy 1977).

After the Gombe and Mahale experience, most other field studies have adopted a more tedious and painstaking process of accustoming chimpanzees to the presence of observers without the incentive of food. The issue was raised again when Boesch (1994a) claimed that hunting at Gombe is also affected by human interference. Although I have found very little corroborating evidence, I have no doubt that there is a seed of truth: the human presence is a biasing factor that can be minimized but not eliminated.

Habituation

To undertake a field study of predator-prey ecology, one must have predators and prey that are habituated to the human presence. This has not been the case for any previous primate study. Habituation can be a tedious and frustrating process that takes many years, but the reward is the ability to spend long hours in close contact with wild animals without influencing their natural pattern of behavior. Fortunately, Gombe chimpanzees are very habituated by virtue of 36 years of close contact with human observers. Researchers maintain a 5-meter distance, but the chimpanzees ignore this rule frequently. On my very first chimpanzee follow, I arrived with field assistants at the night-nest site before dawn, and in the darkness put my hand onto a large rock for support in feeling my way through the dimly lit forest. The "rock" was a still-drowsy chimpanzee just out of his night nest; he flinched but was fairly composed at the contact. Walking along a hilltop ridge with a party of chimpanzees strung out along a narrow trail, I tended to follow at the rear of the procession. Inevitably some chimpanzees lagged behind and I would find myself in the middle of an excursion in which everyone but me seemed to know where we were headed.

Gombe red colobus, on the other hand, were unhabituated and rather flighty animals. While many other monkey species can be easily habituated, red colobus at Gombe do not adjust readily to a human standing underneath their tree; changing this reaction took much time and patience. Why colobus should continue to be afraid of people after seeing them in the forest at Gombe for so many years has been a subject of speculation. They show a fear response to humans similar to that shown to approaching chimpanzees. Is this because they are unable to distinguish humans from their chimpanzee predators?

I was able to test this thesis in 1991, once my two primate study groups were somewhat habituated. As I sat on the bank of Kakombe Stream watching one of my colobus groups, the monkeys burst into alarm calls and started to leap in agitation. I turned and saw that a lone male chimpanzee, Wilkie, had walked up behind me and was staring over my shoulder up into the tree crown. This episode indicated to me both that colobus do distinguish between humans and chimpanzees and that at that point the colobus had become habituated to me. A second possibility is that red colobus, at least at Gombe, instinctively respond with a generalized fear of large-bodied primates, because of their long history of being hunted by chimpanzees. And a third possibility is that because of an ancient history of humans eating colobus in western Tanzania (perhaps thousands or even millions of years ago), natural selection has produced this response—though no modern tribe in that region hunts primates of any kind.

Goodall (1986) wondered if the colobus fright response to humans might not be the result of decades of seeing researchers following chimpanzees; they might have learned to associate the sight of humans with the likely approach of chimpanzees. This seems unlikely in that colobus everywhere in Gombe, whether they have had experience with humans or not, respond in much the same way to people passing under their trees.

Habituation of two colobus groups in Kakombe Valley followed months of being near them and behaving fairly stealthily when I was. They quickly stopped alarm calling or staring at me, though even after a year sudden movements or noises would provoke alarm calling. Since I was interested in part in the vocal response of colobus to chimpanzee approaches, it was crucial to keep them calm enough in my presence to avoid biasing this portion of the study. After the initial research periods, when I returned to Gombe after long absences to gather demographic information, many members of these two groups had lost their conditioning to me and responded as frightened, unhabituated animals. Even early on, a habituated group could spend a morning feeding calmly overhead, then after resting for several hours, waken and burst into alarm calls when they spotted me still sitting beneath them.

These were not the only two red colobus groups that used Kakombe. At least two other groups routinely did, and I saw other colobus occasionally that were either unrecognized groups or fragmented subgroups lacking distinctive individuals who would aid me in identifying the groups.

Observation Techniques

As I have said, my principal goal was to observe the colobus-chimpanzee relationship from the dual viewpoints of the predator and the prey. This sort of study had never been made for any large mammalian predator-prey system owing to the difficulties of observing both species if they were not habituated. It would have been far easier to tape record chimpanzee or eagle vocalizations, then play them to red colobus in controlled settings and record their responses. But such episodes would have been less informative, and the experimental gains would have come at the cost of artificiality. It is also impossible to know with certainty whether colobus respond to cues other than vocalizations when they react to chimpanzees and other predators. Thus they might not perceive the artificial encounters as real. A taped playback accompanied by a life-like model might reliably deceive them.

Without controls based on a sample of true encounters, I considered playbacks less interesting and informative than actual events. Playbacks of chimpanzees could not have been used in any case, as they might have elicited a territorial response from any chimpanzee within earshot. Recordings have been used with fascinating results in other field sites to study primate reactions to predators (Cheney and Seyfarth 1991; Noë 1996; Treves, unpublished). In these cases a field study of actual prey-predator encounters was impossible because of the unpredictable habits of the predators.

Nor would playbacks have allowed me to study the second of my interests, the hunt itself. Why do chimpanzees engage in such a major expenditure of time and energy for a seemingly small return, and what factors lead to success or failure for both sides? To address any of these questions, actual encounters between the two species were necessary.

Major logistical problems arose, however, in observing encounters that might lead to hunts. Chimpanzees travel far and wide in unpredictable small subgroupings, not in easily located and followed large, stable groups. On a given day the males, who do most of the hunting and therefore were the main objects of my research, could be in any of eight valleys in groups up to 35 animals. I was warned by one primatologist before setting out that it would be impossible to see much hunting by watching the prey, any more than one could study the hunting behavior of lions by following one zebra all day long. Chimpanzees scatter far and wide and may attack any of dozens of equally scattered colobus groups. Researchers interested in hunting have therefore always followed the hunters. Furthermore, although hunting is a routine behavior for wild chimpanzees, it is still fairly infrequent. Hunts occur about twice a week in an average year, but with tremendous variation—including (although it was not known when I began) variability by season and by month.

The solution to these problems came about gradually in the course of spending time in the field, learning how the animals spend their days, and reading thousands of pages of Gombe Research Center field notes recorded in Kiswahili* by the Tanzanian field assistants. The topography of Gombe was also instrumental. Because of its hills, Gombe provides a researcher with a benefit many other chimpanzee habitats lack—vantage points. The Peak, famous as Goodall's locale for watching unhabituated chimpanzees in the early months of her work, served well as a lookout. In the early morning, chimpanzees who had nested in Kakombe Valley or high on the slopes above it could be heard

* Kiswahili is the correct name for the language common to the East African countries, usually referred to as "Swahili."

giving choruses of pant-hoots as they woke and traveled in search of breakfast and other chimpanzees. Each ravine within the valley is mapped and numbered; once a colobus group had been sighted in, for example, KK5 (the fifth Kakombe korongo, counterclockwise from the southwestern corner), it was usually not difficult to make my way to that area and find the group.

This protocol, combined with a day-to-day knowledge of where colobus study groups had been feeding, made it possible to track a chimpanzee party from high above Kakombe. The gamble was to choose the most likely group and the most likely location in which an encounter would occur. I would then hike or climb rapidly to arrive at the colobus group and record encounter data before the chimpanzees approached. I chose correctly about half the time. Knowing what food patches both chimpanzees and colobus were using in each season helped a great deal in predicting travel patterns.

I probably was able to see about three-fourths of the hunts that took place in Kakombe during the time I was in the field (extrapolated from the estimated total hunts during that time), while spending many other days collecting information on the colobus themselves. In practice, there were many days on which, after hours alone with the colobus, I would hear chimpanzee pant-hoots, and as the chimpanzees approached I collected data on colobus behavioral response. Most often the pant-hoots eventually ceased or changed direction. When they did not, an encounter took place.

I wanted to follow the chimpanzees also, to understand better why they hunted and what factors influenced when, where, and how they did so. That was not my original plan, but it emerged after my first few months in the field and became a central part of the research. One can spend weeks following male chimpanzees, however, and see little or no hunting. I needed to learn if there were patterns of hunting that would enable me to predict when and where hunts were likeliest to occur, in order to maximize my time with the chimpanzees and still leave enough hours to gather information about the colobus.

Each day of the year a target chimpanzee is followed by a pair of Tanzanian field assistants, led by Hilali Matama. By compiling and analyzing hunting ecology from field notes through the 1980s (Stanford et al. 1994b), I was able to draw a clearer picture of Gombe chimpanzee hunting patterns than had been available previously. It became apparent that Gombe chimpanzees hunted much more often when in large parties than when only a few traveled together, and that the dry months of July–October were the peak hunting months.

Armed with this new information and with the help of the field assistants, I was able to plan each research day according to information obtained the previous night: where the chimpanzees had slept and who had nested with whom. I particularly followed parties in which the adolescent male Frodo was present, since he was the most avid hunter and a catalyst for the others. In addition, knowing where each colobus had last been seen in Kakombe Valley, I judged whether I was more likely to see an encounter the next day by following the chimpanzee party or by relocating a colobus group. Data in this book were collected by both methods and present two different views of the hunt, from the standpoint of the attackers and of the victims. Knowing that a hunting peak occurred at the middle and end of the dry season was also essential, since I knew when to arrive at Gombe to hope to see more hunting in a relatively short time. Using this strategy, I saw a fairly constant average of one hunt and two chimpanzee-colobus encounters per week in the field, in addition to obtaining the separate data that I was collecting on each species.

Data Collection

All told, I spent about 21 months at Gombe over a five-year period (May–December 1991, June–November 1992, July–August 1993, August–October 1994, and July–August 1995). Once it became clear that hunting peaked strongly in the dry season and was rare in the wettest months, I conducted field sessions during the dry-season months. This approach maximized hunting data, though as a result I was not at Gombe to record complete annual measures of diet and ranging of the colobus. I also missed some colobus births. That lack of information would have constituted a critical gap in understanding the effects of chimpanzee predation were it not for the research projects on Gombe colobus socioecology of two doctoral students: Sharon Watt of the University of Auckland, New Zealand, and Shadrack Kamenya of the University of Colorado, Boulder. They filled the gaps in my own collection of data on predation by studying colobus ecology and birth seasonality during the same period (Appendix 3).

Splitting my time between two species, I was never able to compile a thorough natural history of either. In the case of the chimpanzees, this was unimportant because of the quantity of data collected at Gombe over so many years. I spent an estimated 2,000 contact hours with red colobus, and recorded their behavior systematically for about 1,400 hours (Appendix 4). I spent approximately 650 hours with chimpanzees, the majority of that time with medium to large parties that contained at least several adult males. Following lone males is not a very productive way to study hunting; they rarely hunt, and when they do are usually unsuccessful. Lone males tend to locate other males and they often led me to larger parties. Because one particular male, Frodo, was such an avid hunter and (contrary to pattern) hunted alone frequently, I followed him when he was alone more often than other solitary males.

I collected data of three different sorts on the colobus. In 1991 I spent most of my time following two red colobus groups in and around Kakombe, to habituate them and to collect basic ecological and behavioral information. These data were recorded on sheets modified somewhat from my previous work with another colobine monkey, the Asian capped langur (*Presbytis pileata,* Stanford 1991a). I recorded information on location, group spread (distance between the two individuals farthest apart), nearest neighbor (distance between group members), foods eaten, activity engaged in, and all forms of social behavior: vocalizing, grooming, copulating, threatening or fighting, and playing. By measuring what colobus did with their time, how high in the forest they foraged, how closely they clustered when feeding or resting, and how far they traveled in a day, I established a baseline against which to record variations in the same behaviors just before, during, and after a hunt. I continued this routine through 1992 and identified all of the adult and many of the immature animals in the two groups, called J and W.

During this time I learned to identify three other colobus groups in the Kakombe Valley area (MK, AK, and C), based on the distinguishing marks, scars, and tail deformities of one or two adults in each group. I then followed these five groups for five years, despite the difficulty of reliably censusing the groups. Finding and following one of my three monitored red colobus groups in the Kakombe Valley for an entire day would not guarantee seeing every group member because of the dense foliage, the terrain, and the tendency for the group to forage while dispersed over areas of 50–100 meters or more. These unhabituated study groups were often counted from across a ravine, where I could watch them from a distance with a spotting telescope as I made notes. If the group had not been seen in weeks or months, I noted additions and absences, which often meant following for several days. I attempted to find, follow, count, and recount group members at least three times each month until I felt sure I had seen all animals in the groups. One of the two main groups, W, dispersed in 1994.

The second category of information that I collected addressed a key question: how do red colobus respond to the imminent threat of being attacked? My sample consisted of 119 encounters, more than half of which were observed while I was with the colobus as chimpanzee parties approached.

I used different methods to obtain information on these encounters. The auditory aspect of the hunt was important data that I wanted to obtain. I initially recorded events on a Sony Pro-Walkman tape recorder, pointing a Sennheiser M-80 microphone at the colobus while narrating into a small lapel mike. The tape ran in real time when an encounter appeared to be imminent, a technique that produced many valuable samples of colobus calls in response to predation threats. If a hunt did begin, however, the ensuing minutes spent scrambling through dense thickets often left my equipment in no shape to record the happenings overhead. By 1993 I had switched to the less innovative but more reliable notepad and only sometimes recorded the vocalizations. In either case, I recorded colobus behavior that occurred before the chimpanzees had been seen: changes in spatial patterns within the group, changes in the frequency and type of vocalizations, and changes in the patterns of male association. The defense tactics, number of males defending, and roles of other colobus present were recorded as well.

A video camera would seem to be ideal for recording a complicated group behavior sequence like a hunt, since the tape can be replayed later to analyze the action. Unfortunately, watching through a video camera gave me a uselessly narrow view of the event. In addition, watching a hunt in the treetops through the tiny camera monitor rendered the action nearly invisible. Having a field assistant operate the camera while I watched would have introduced an additional person into the scene, something I was reluctant to do. In the end I gave up using the video camera and instead took many still photos. Most hunts that I saw were against unhabituated red colobus groups; but when one of the study groups was attacked, which happened eight times, I was able to observe and record the response of specific individuals. I also recorded information about the microhabitat in which the hunt occurred, the composition of the chimpanzee party, and the food patch in which the chimpanzees had been feeding or traveling.

Encounters were counted if chimpanzees passed within 50 meters of a colobus group overhead or to the side of the travel path. Colobus groups were often heard at greater distances when I was following chimpanzee parties. The chimpanzees presumably heard these calls, but they were not considered encounters. A hunt began when one or more chimpanzees ascended trees holding or adjacent to red colobus groups. Often only one or two chimpanzees hunted while the rest watched; this was still scored as a hunt.

In calculating the effects of party size and number of males on the outcome of hunting, I used as the hunting party size the number in the overall party, rather than just the number of males actively pursuing (because party size and number of males were the most robust factors affecting a hunt's outcome). I learned from watching many hunts that chimpanzees who at first appeared to be passive bystanders often played key roles as the hunt progressed.

I therefore used a criterion different from that utilized at other sites: "hunters" were only those chimpanzees actively pursuing prey. When a chimpanzee party decided to hunt, I recorded the sequence of events, noting the roles taken by each hunter and also by the "bystanders." I registered any apparent cooperation and how it might have affected the outcome, and attempted to see the hunt as a sequence of actions of individuals as well as a group effort at catching a colobus.

Although I observed a small number of hunts of other mammal species (see Chapter 4), only encounters and hunts involving red colobus are analyzed in this book. The reason for this exclusion is obvious to anyone who has compared the hunting styles of chimpanzees as they pursue various prey species. Only in hunting colobus do the chimpanzees face a prey animal that stands its ground and mounts a group defense that can affect the outcome. Captures of the other species are highly fortuitous. A baby pig or bushbuck is stumbled upon in the forest undergrowth, whereupon the chimpanzees grab it and flee from the enraged parents. The hunts of baboons and blue monkey that I have seen are also rapid pursuits of fleeing animals. Planning is apparent only in the hunting of red colobus, and it is on this basis that a number of provocative questions can be asked.

3. The Hunters

24 November 1991: In late morning a party of the Kasakela chimpanzees is on Sleeping Buffalo, the wooded ridge that separates Kakombe Valley from Mkenke Valley. More than 20 members of the community are feeding and socializing, with more animals arriving every few minutes. Spray, a female with a pink estrous swelling, is present and a fight breaks out when she declines to copulate with Goblin as promptly as he would like. He chases her into a tree crown where she fear-grimaces and then allows him to mate with her. Mothers and infants are scattered along the trail below the trees in grooming clusters and play groups.

At midday the males abruptly depart toward the south, leaving most of the females and young behind. Wilkie, Freud, Frodo, Goblin, Beethoven, Atlas, Apollo, Gimble, and Tubi walk in loose formation along trails and through thickets until they reach the edge of Kahama Valley. There they climb a tall tree that overlooks Kahama and stare vigilantly into the southern border region of their territory. Eventually they climb down and continue, though more slowly and in single file. No one makes a sound, and occasionally the entire party freezes in midstep to listen to some far-off noise before proceeding. When I step on a branch and snap it loudly, a dozen faces glare back at me.

A patrol like this one can last for hours and take a party of males to their territorial perimeter and beyond, into the no-chimpanzee zone that is used by two adjacent communities and in which both parties travel with extreme caution. As the party reaches Kahama Stream, field assistants Hamisi Mkono and Msafiri Katoto and I expect the males to turn back. But they continue, crossing into the territory of the adjacent chimpanzee community. Now there is no vocalizing and no grooming. The party heads into Nyasanga Valley. The members stop at a termite mound, but instead of fishing for the insects they inspect the termiting sticks left by the previous termiters.

They travel until they are a half-kilometer inside enemy territory; Mkono says this is the farthest south a Kasakela patrol has penetrated in many years. Finally the patrol reaches the beach near Nyasanga Stream, where they stop to termite at a mound strewn with the tools of the southern chimpanzees. Tubi sniffs tree trunks in the vicinity, apparently trying to detect the recent presence of the residents. They have not heard pant-hoots, and in late afternoon they turn and head north on a ridge overlooking the beach until they have crossed Kahama Stream. At dusk they build their night nests well within their territorial range in Kahama Valley.

During this territorial incursion, the males have patrolled the border of their territory, monitoring it for unknown chimpanzees, and obtained information on the whereabouts of food sources in the territory of the alien community.

Our understanding of chimpanzee society emerged slowly because of the fluidity and complexity involved. Moreover, chimpanzee life stages such as gestation, infancy, and adolescence are nearly as long as those of humans. Collecting data adequate to understand long-term patterns of behavior therefore takes decades. The social structure of the chimpanzee consists of an intensely territorial community of related males, plus females who migrate between communities after puberty. Within this community structure individuals meet and travel together for hours or days, parting unpredictably.

The label "fission-fusion polygyny" is often used to describe chimpanzee society. The exception is mothers and their dependent offspring, who form the only stable social unit. Females are not highly sociable when they are not sexually swollen and spend much time foraging alone. The temporary subgroupings, or parties, vary in size according to two main factors: the availability and the size of ripe fruit patches and the availability of estrous females. Both ripe fruit and estrous females are targets of attention by males, and both are only ephemerally available. The size of the party can therefore vary between one and several dozen depending on the day and the season.

At Gombe the average party size is four to six chimpanzees; this average can be much higher in years when many of the popular adult females are reproductively fertile (Goodall 1986) or in periods of fruit abundance (Wrangham 1975). At such times the females become highly social members of the community, and large parties traverse their range from one favored fruit tree to the next.

Because they remain in their natal territory throughout their lives, males tend to be related to one another (Morin et al. 1993). This kinship promotes the formation of male coalitions that enable individuals to rise in the dominance hierarchy or control access to females.

Males often travel in groups to patrol the perimeter of their territory for incursions by strangers. During patrols males greet young fertile females with great eagerness but brutally attack noncycling females (Goodall et al. 1979). In the most recent such episode at Gombe, a patrol of Kasakela males ambushed a northern (Mitumba) community female with her newborn baby and a son. The female and her older offspring escaped but the males took the female's infant from her and killed it. Such events are infrequent but may still account for the death of many members of the community over the long term.

At least twice during the history of human observation of wild chimpanzees, the males of one community have systematically and savagely attacked, one by one, the members of an adjacent community until the other community was destroyed. This warfare occurred at Gombe over several years during the early and mid-1970s (Goodall et al. 1979). Shortly thereafter, a similar intergroup conflict took place at Mahale, in which M-group systematically eliminated adjacent K-group (Nishida et al. 1985). Although the explanation for these events is not fully understood, the Gombe community had a large complement of adult males, who slowly fissioned along alliance lines until the community had effectively split. Eventually parties of each group began to appear hostile toward parties of the other group.

The flexible grouping pattern of wild chimpanzees provides primatologists with a rich array of questions about primate sociality. The primate order includes species that live in large groups, small groups, one-male groups, and multimale groups. Chimpanzee parties may temporarily resemble any of these social structures under particular circumstances of female or fruit availability. When larger parties form owing to food abundance or swollen females, or for patrols, hunting is much more likely to occur than at any other time (Stanford et al. 1994b). This allows testing of hypotheses about the value of party size and composition in hunting success.

Males and Females

The relationship between male and female chimpanzees is one of dominance, power, and manipulation. Male chimpanzees begin to rise in the dominance structure of the community upon reaching adolescence. They start by dominating the adult females, each of whom is typically subordinate to the lowest-ranking adult male (Goodall 1986). This process can be gradual or quite rapid. By challenging—through charging displays and outright attacks—all members of the community whom he is able to contest, a male determines his eventual position in the male dominance hierarchy.

Many observers have considered chimpanzee society to be male dominated and regarded females as powerless in the face of male aggression and sexual coercion. Females should not, however, be considered passive. Some females become high-ranking community members, even power brokers among males. Females are also active solicitors of matings and are famous for their promiscuity. Caroline Tutin (1979) recorded females copulating with as many as eight males in a five-minute period during her study of the reproductive strategies of Gombe female chimpanzees. Gagneux et al. (1997) showed that among Taï forest chimpanzees, females may live in one community but conceive offspring with males of other communities as well.

Female mating overtures are often preceded by cues from the males, such as branch shaking or specific hand or body gestures (Goodall 1986). If a female fails to respond to such signals to the male with sexual swelling presented, she may face the wrath of his frustrated attack. For example, a male may sit near an estrous female, reaching out to shake a bush or branch every few minutes until the female approaches and offers her swelling for mating. If she does not respond, the male is likely to become impatient and the female may become a target of his frustrated aggression.

Adult male chimpanzees also pursue a reproductive alternative to polygynous mating. This is the consortship. Although other primate males engage in some sort of temporary sequestration of females during their periovulatory periods (baboons for example, Hausfater 1975), these segregations are usually of shorter duration than in chimpanzees. Some male chimpanzees coerce females to follow them to areas of the community range far from the core area. They then endeavor to keep these females away for days, weeks, or even months. The end result of these sequestered couplings is that a large percentage of babies are conceived by females while alone with the consorting male (Tutin 1979). Older, past-prime males in particular use this reproductive tactic. Goblin, the former alpha male, frequently led females to a northern valley, remaining with them through one or more monthly estrous swellings. Even alpha males will temporarily give up their central position in the community to go off on a consortship. The alpha males of the Kasakela and Mitumba communities, Freud and Cusano, each disappeared in the company of females and stayed away for weeks during 1995, only to return later and resume command.

When a female chimpanzee migrates, she enters a new community in which the resident females are largely unrelated to her. There is thus little genetic incentive for cooperation among females, and relatively little affiliation occurs. Even when several females and their offspring sit in proximity, the females will not necessarily engage in reciprocal grooming or support one another if tensions erupt. Females form coalitions far less often than males cooperate. Close bonds exist between some males, and between some males and females, but rarely between two unrelated females (Goodall 1986; Nishida 1990).

Chimpanzee Hunters

Among the twelve adult and adolescent males at the start of the study were four pairs of brothers, plus Wilkie (Table 3.1). Wilkie, the alpha-ranking male from 1989 through early 1993, was a small, extremely muscular male. Most of our impressions of what an adult male chimpanzee should look like are formed by zoo animals, who grow quickly and to large size because of their rich diet. But Wilkie was no bigger than many captive adolescents. Gombe chimpanzees tend to be smaller than those at other study sites (Morbeck and Zihlmann 1989) perhaps because of the seasonal dryness, and Wilkie was small even by Gombe standards. He weighed about 40 kilograms, though as alpha his hair was bristled almost constantly, giving him an impressive posture. Wilkie had risen to power in a fight with longtime alpha Goblin that began over an estrous female. He ruled confidently with guile and political savvy rather than by brute force. Wilkie invested much time and energy in currying the favor of the high-ranking female Fifi and was often accompanied by his ally Prof.

After losing the alpha rank to Freud in early 1993, Wilkie took on an

TABLE 3.1. *The males of the Kasakela chimpanzee community, 1990–1995.*

NAME	CODE	AGE IN 1995	KNOWN ADULT KIN IN KASAKELA COMMUNITY	RELATIVE RANK DURING STUDY PERIOD
Evered	EV	Died 1993 at estimated 41	Father of Wilkie	Intermediate
Goblin	GB	31	Brother of Gimble, Gremlin	High
Atlas	AL	28	Brother of Apollo	Low
Beethoven	BE	Estimated 26		Low
Freud[a]	FD	24	Brother of Frodo, Fanni, and Flossi, son of Fifi	High (alpha 1993–)
Prof	PF	24	Brother of Pax	Intermediate/low
Wilkie	WL	23	Son of Evered	High (alpha 1989–1993) to intermediate/low
Frodo	FR	19	Brother of Freud, Fanni and Flossi, son of Fifi	Low to high
Gimble	GL	18	Brother of Goblin and Gremlin	Low to intermediate
Tubi	TB	18		Low
Pax	PX	18	Brother of Prof	Low
Apollo	AO	16	Brother of Atlas	Low

a. Freud was replaced as alpha by his younger brother Frodo in late 1997.

entirely different appearance. Without his hair constantly piloerected, his diminutive size was apparent. Wilkie plummeted in rank once he had been deposed, traveling alone for a time, but by 1995 he returned as a low-ranking member of the community. He had few close relatives, though DNA testing later showed that he was the son of the elderly male Evered (Morin et al. 1993). Wilkie provided proof that being a skilled hunter does not necessarily correlate with high rank; he was afraid of male colobus and often retreated at the first sign of their approach.

The male whom Wilkie deposed in 1989 was Goblin, who had reigned as a powerful alpha for most of his adult life. In the early 1990s he was reentering the social life of the community after a period of ostracism following his overthrow. Goblin was also a small male who had ruled by political astuteness, and though he was no longer top-ranking he was again a powerful partner to males with whom he allied himself. His power in the community lay in his tendency to shift alliances from day to day, changing the balance of power in the community accordingly.

Goblin's younger brother Gimble, one of a pair of twins born in the 1970s, was an adolescent male who began a rapid rise up the dominance hierarchy in 1994. By 1996 he had become the fourth-ranking male in the community behind Freud, Frodo, and Goblin. Goblin's sister Gremlin was one of the most important adult females in the community as well.

I became interested in Goblin as a hunter after I had examined 10 years of hunting data from the 1980s. I discovered that while Goblin was alpha he had captured or stolen an extraordinary number of colobus monkeys. Was this an effect of being the alpha or a reflection of his hunting skill? Watching Goblin hunt with the other males now that he was no longer dominant might answer this question.

The brothers Freud and Frodo were the most interesting pair of males of the Kasakela community. They were the latest adult males in the powerful F-lineage reported on by Goodall since the early 1960s, beginning with Flo and her sons Faben and Figan. The first and second sons of Fifi and the grandsons of long-deceased Flo, Freud and Frodo had long been projected by Gombe researchers as future leaders of the community. By 1991 this prediction appeared ready to be fulfilled. Both were large, handsome chimpanzees (though they did not resemble each other) with entirely different personalities.

Freud was 20 years old as my study began, an age at which males often rise to alpha status. Although he was clearly high ranking, Freud seemed completely uninterested in the politics of the male dominance hierarchy. His demeanor was nearly always placid and he was frequently alone. His calm nature notwithstanding, he rarely gave the submissive pant-grunt vocalization to Wilkie, and received frequent charging displays from the alpha as a result. Most of the Gombe researchers were surprised when, in early 1993, Freud challenged Wilkie and over a period of weeks became the alpha male of the community. He still held this status in early 1997. Because Freud was a powerful male and had a brother in the community, he and Frodo provided an opportunity to examine the kin-based benefits of cooperation in hunting and meat-sharing.

Frodo was a 14-year-old adolescent when I began my study, and I watched him grow over the years into an imposing adult male. By 1995 he was the second-ranking male behind his elder brother, Freud. Frodo was noteworthy in part for his size. Even as a 14-year-old he was the largest male in the community, dwarfing Wilkie and Goblin. By age 15 he weighed more than 50 kilograms; his aggressive and intimidating demeanor was felt by the colobus, other chimpanzees, and his human observers as well. When displaying he charged into, rather than around, researchers and brutalized the easily dominated younger chimpanzees in the community. He was nicknamed *jambazi* (the rogue) by the Tanzanian researchers for his tendency to hit, kick, and charge them. He was also called the *mwindaji fundi* (expert hunter) for his fearless pursuit of colobus and his willingness to withstand counterattacks by male colobus that drove off all the other Kasakela males.

Frodo's physical size and physique were not, however, well matched with his social or political status. He seemed uninterested in currying the favor of males who might have been useful to him politically. As he began to rise in rank, he was continually rebuffed by Wilkie and later by Freud, both of whom obviously viewed Frodo as a threat. By 1995 he had matured into a high-ranking male who nevertheless had few allies and whose obstacle for top rank was his dominant elder brother.

Evered was the only male chimpanzee remaining from Goodall's earliest days of research. In 1991 he was estimated to be 41 years old and showed signs of advanced age. His teeth had been lost or worn down; only one broken canine was visible in his mouth. Even without the weaponry or the agility often associated with hunting prowess, Evered remained a highly successful hunter until his death. This is evidence of the importance to successful hunting of coordinated effort and experience rather than large canine teeth or arboreal adaptation. DNA tests showed that he was Wilkie's father (Morin et al. 1993), though the lack of paternal caregiving by male chimpanzees makes it unlikely that he recognized this kinship. Evered was one of Wilkie's main allies during the latter's tenure as alpha.

Prof was Wilkie's other close ally. He was a highly successful hunter whose status rose while Wilkie was top ranking but fell after Wilkie had been deposed by Freud. Prof supported Wilkie and in exchange received Wilkie's favors. While Gremlin was swollen (and presumably ovulating) in 1992, she spent six consecutive days at the peak of her swelling in the crowns of trees being mated and guarded by Wilkie. As Gremlin foraged from tree crown to tree crown, Wilkie followed closely while the other adult males watched their movements from the ground beneath. Any male who approached Gremlin was chased off by Wilkie. Only Prof was allowed to copulate with Gremlin.

This episode illustrates that chimpanzee society should not be characterized as promiscuous. While females engage in a multiple-male mating strategy, males often compete for swollen females. Females vary greatly in their attractiveness to males. Several females at Gombe—Fifi, Gremlin, and Patti—become the foci of huge foraging parties and create much excited competition among the males when they are swollen. Both major dominance upheavals among the Gombe males were triggered by competition over swollen females: Wilkie overthrowing Goblin in 1989 (Candy), and Freud overthrowing Wilkie in 1993 (Patti). Other females generate only mild interest among most of the males and are more often mated by younger or lower-ranking males.

Beethoven was another adult male frequently present in hunting parties. This low-ranking chimpanzee was often accompanied by an orphaned female, Dilly, of whom he was quite solicitous. Beethoven entered the community as the infant of an immigrant female, Harmony, and was the only Kasakela male not born as a community member and without known kinship ties there. Atlas, the elder brother of Apollo, was also a low-ranking male. Prof's younger brother was Pax, who had been orphaned at an early age when his mother Passion (notorious for her cannibalistic tendencies), died. As a result, Pax's size was that of a juvenile; though he was a frequent hunt participant, he rarely was able to keep his catch away from the larger males. Tubi and Apollo were maturing adolescents eager to join in the male patrols and hunts.

Adult and adolescent males were not the only hunters in the community. Female chimpanzees relish meat as much as males do, and when the opportunity arises they capture prey. Goodall (1986) suggested that female participation in hunting was greater during the 1980s than was generally appreciated, but Gombe data for past two decades indicate the opposite; females account for a very small percentage of kills.

Gigi was the most frequent hunter among the females of the Kasakela community. She was an elderly female who died during my study at an estimated age of 38. She was presumed to be infertile, having had regular swelling cycles throughout her life without producing any offspring. Gigi was a frequent member of hunting parties and a successful hunter herself. She was never burdened with dependent offspring; whether this influenced her active participation in hunts is unknown, but it seems likely. Gigi also possessed a more muscled, male-like body than the other females. Fifi and the mother-daughter pair Sparrow and Sandi were the other adult females who were eager hunters and occasionally caught prey. Gigi accounted for 4 percent of all kills; other females combined accounted for 4.5 percent. All other kills were made by the adult males.

Other Potential Predators

Chimpanzees are only one set of potential hunters of colobus monkeys at Gombe and elsewhere. Unfortunately, little information exists on the other predators, so we must examine studies conducted in other forests. Most information about the diet of big cats, for example, is based on examination of prey remains in the feces rather than on direct observation (Schaller 1967, 1972). The evidence suggests that in each geographic region except Southeast Asia some birds of prey are more important predators on wild primates than the big cats are.

Predators on primates at Gombe and in other tropical forests fall into several classes, each of which presents a distinct type of predation risk and style of hunting.

BIRDS OF PREY. The best-documented nonprimate predators on primates are raptors. These can be either diurnal (eagles and hawks) or nocturnal (owls). Birds of prey may attack while soaring above the forest canopy or from within it, and the many raptor species have markedly different strategies of hunting (Brown et al. 1982). The threat posed to wild primates by raptors is a function of the relative body size of the primates and the raptors, the feeding behavior of the raptors, and the density and distribution of raptors in ecosystems that have primate populations. In the neotropics, for instance, the squirrel-sized monkeys of the family Callitrichidae are potential prey not only for the massive harpy eagle *(Harpia harpyja)* but also for smaller eagles and hawks, several species of which inhabit most primate territories. Based on this predator size/prey size correlation, the largest New World monkeys—the spider monkeys *(Ateles),* the howlers *(Alouatta)* and the muriqui *(Brachyteles)*—should be less vulnerable to predators than the marmosets and tamarins (Terborgh 1983). Goodman et al. (1993) interpreted antipredator behavior by medium-sized Malagasy lemurs as an evolved response to a raptor that is extinct on the island.

Across Africa the raptor most often reported as a predator on wild primates is the crowned eagle, which preys on animals weighing up to 30 kilograms, or six times its own weight (Daneel 1979). Most prey are smaller, but this raptor is able to capture adults of most African arboreal monkey species. The martial eagle *(Polymaetus bellicosus)* is the other large African eagle, though it inhabits open country and is less often implicated as a significant Leslie Brown and colleagues (1982) reported monkeys as an important part of the crowned eagle diet in Kenya. This species preys on baboons in Gombe National Park (Gombe Stream Research Center, unpublished), and on other monkeys in Kibale National Park (Struhsaker and Leakey 1990) and in Taï National Park (Boesch, personal communication). In some areas its diet may be almost entirely monkeys (Steyn 1982).

Unfortunately, most of our information about crowned eagles comes from open country or broken woodland, while most primate populations on which the eagle preys live in densely forested habitats. While most raptors hunt alone, some species hunt in pairs that may cooperate to increase the chances of a kill. The crowned eagle is one of the species reported to be a cooperative hunter. When hunting one crowned eagle may sit quietly in the forest watching for prey and attack from beneath the canopy. A hunting partner meanwhile attacks from above, perhaps better able to capture monkeys that are concentrating on the other eagle beneath (Leland and Struhsaker 1993; Boesch, personal communication).

The crowned eagle's impact on monkey populations is well documented. Joseph Skorupa conducted a study of crowned eagle breeding biology and diet in Kibale National Park, which has one of the highest primate biomasses of any tropical forest. During one breeding season he monitored an eagle nest and documented the remains of prey either brought by the eagles or found on the ground nearby. Nearly 88 percent consisted of primates (the remainder were duiker antelopes). Of the primate kills, the species most represented was black-and-white colobus (39 percent of the total; Skorupa 1989).

In another study of eagle predation in Kibale, Struhsaker and Leakey (1990) collected prey remains over a 3-year period and correlated their information with primate density information obtained over many years. Monkeys were again the predominant prey species (83.7 percent). Several interesting findings emerged from this study. First, black-and-white colobus and red colobus accounted for a combined 40.5 percent of all prey eaten. Second, the proportion of red colobus who were immatures was significantly greater than expected from their numbers in the population; eagles selectively preyed on young red colobus while selectively preying on adult black-and-white colobus. Struhsaker and Leakey suggested that because red colobus at Kibale live in multimale groups that aggressively defend against potential predators, adults rarely fall prey to eagles. One-male black-and-white colobus groups incur greater losses of adults because their social system leaves adults more vulnerable.

The tendency of all species other than black-and-white colobus to form mixed-species groups may offer predator protection to those species. Black-and-white colobus may not form mixed-species associations because of their larger body size. Struhsaker and Leakey's study supports the observation that raptors are more likely to be selective in choosing primate prey than cats are, and that the overall take of primate prey biomass by raptors is higher than recorded for any felid that has been studied.

The harpy eagle is the largest raptor of the neotropics and is also the world's most powerful bird of prey. Now rare as a result of loss of the pristine forests in which it lives, it is a heavy predator on wild primate populations in Central and South America. Neil Rettig studied the biology of harpy eagles in Guyanan rain forest in the 1970s. He found that of 57 prey items brought to the nest (mainly by the males), 18 (31.6 percent) were monkeys, especially *Cebus* ssp. Of a total estimated prey biomass of 38.8 kilograms brought by the eagles, 16 kilograms (41.2 percent) were primate biomass (Rettig 1978). In a related finding, Robert Izor (1985) found that nearly all of the prey remains under the nest of a breeding pair of harpy eagles in Guyana were adult primates.

Nocturnal birds of prey are also potentially important predators on primates. In fact, the most systematic information about predation on a prosimian population comes from a study of Madagascar barn owls *(Tyto alba)* and long-eared owls *(Asio madagascariensis)*, both medium-sized species that occur in various habitats across Madagascar. Steven Goodman and his colleagues (1993) estimated that 25 percent of the local population of mouse lemurs *(Microcebus rufus)* in Beza Mahafaly Reserve was taken by owls, based on analysis of owl prey remains and census data of the mouse lemur population. Mouse lemurs are the smallest primates, so we expect them to be highly vulnerable to predators of many types, even on Madagascar where the diversity of predators is low. *Microcebus* breeds rapidly, reaching sexual maturity in 8 to 12 months and thereafter producing at least one and sometimes two litters of two or three offspring per year (Tattersall 1982). Goodman et al. (1993) speculate that this high reproductive rate diminishes the impact of predation on mouse lemur populations.

FELIDS AND OTHER CARNIVORES. The big cats are major predators on many wild primate populations. In his field study of tigers and their ungulate prey species in central India, Schaller (1967) found that 7 percent of all tiger scats contained hair of Hanuman langurs *(Presbytis entellus)*. Without information on langur population size and tiger ranging habits, we cannot extrapolate this figure to the percentage of the langur population eaten. The finding is nonetheless important because most researchers had assumed that leopards were the principal predators on langurs, while tigers specialized in larger prey.

It appears that for the large Asian cats, the abundance of alternative prey determines whether they include monkeys in the diet (Seidensticker 1983; Seidensticker and McDougal 1993). Leopards are assumed to prey on African and Asian monkeys, but their diet is less known than that of the other big cats. In Taï National Park, Boesch (1991) recorded leopard predation on his chimpanzee community frequently enough to suggest it may be a factor influencing chimpanzee grouping patterns. A study of Taï leopard scats by Hoppe-Dominik (1984) indicated that leopards do not select primates as prey over other equally available mammals. In South Africa, Mills and Biggs (1993) found that 10 percent of leopard prey consisted of primates, mainly chacma baboons. Even the largest primates are not immune; Michael Fay and his colleagues (1995) found gorilla parts in leopard dung in the Ndoki Forest of Central Africa.

Lions are the largest cats in any African ecosystem containing primates and are known to prey on primates. Most reports are anecdotal, such as predation on baboons in Tana River National Primate Reserve in Kenya (Condit and Smith 1994). In the Mahale Mountains of Tanzania, Tsukahara (1993) documented chimpanzee remains in lion feces that coincided with the appearance of two lions in the study area, plus the disappearance of four adult and adolescent chimpanzees. The lions had moved into the study site after the departure of the local human population, suggesting that in more pristine habitats lions might be regular predators on chimpanzees and other large-bodied primates.

New World cats include two large species, the jaguar (*Panthera onca*) and the puma or mountain lion (*Felis concolor*). The puma is an ecologically ubiquitous carnivore that preys on animals as large as deer across most of the Western Hemisphere, including tropical and subtropical primate habitats. There is, however, no record of a puma eating a nonhuman primate. The much smaller ocelot (*Felis pardalis*) preys on small-bodied primates in Peruvian rain forests (Emmons 1987). The jaguar has often been implicated as a primate predator, and at least one record exists of severe depredation of a primate group by jaguar. Peetz et al. (1992) documented the loss of five of six group members of red howlers (*Alouatta seniculus*) from probable jaguar predation over a seven-month period on an island in Venezuela. In Peruvian rain forest, Louise Emmons (1987) found that puma and jaguar appear to be separated by habitat preferences, with the jaguar a riparian habitat forager.

Confounding variables such as predator population density, size of hunting range, feeding frequency, microhabitat preferences, and prey size make it uncertain whether felids are a greater threat to wild primates than birds of prey (Brown and Amadon 1968); the proportion probably differs from one predator-prey relationship to the next.

In one of the most predator-rich primate study sites, Manu National Park in southeastern Peru, primates do not figure prominently in felid diets (Emmons 1987). Dietary overlap is high among puma, jaguar, and traditional foraging people living in the same forests. One notable dietary difference is the abundance of monkeys in the human menu and their scarcity in the diets of the cats (Jorgensen and Redford 1993). Comparable data from African felids are available only from the Taï Forest, and they support Emmons' findings for neotropical felid communities. It does not appear that forest cats specialize in particular prey species, including primates, to the extent that either savanna cat species (Caro 1994, cheetahs), or large raptors do.

A variety of small carnivores may prey on either the smaller species of primates or the immatures of larger species. Few of these carnivores have been well studied, and they may not be frequent predators on primates. In Madagascar, for instance, the only carnivore large enough to pose a threat to the prosimian primates on the island is the fossa (*Cryptoprocta ferox*). A viverrid that resembles a medium-sized cat, the fossa is an important predator on at least some lemur species (Goodman et al. 1993; Wright 1995; Overdorff 1995).

Wild canids are also occasional predators on wild primates. My previous anecdote of golden jackals (*Canis aureus*) hunting capped langur monkeys (*Presbytis pileata*) in Madhupur National Park, Bangladesh, is one example (Stanford 1989). Jackals are common scavengers in Asian and African ecosystems, but have rarely been reported to prey on animals as large as a 10-kilogram monkey. Lambrecht (1978) hypothesized that in habitats where larger predators such lions or tigers were extinct, the smaller social carnivores might occupy their role, preying on larger animals by hunting in groups. This theory may explain jackal behavior in capped langur habitats. Jackals appear to monitor trees in which langurs are feeding, perhaps in hopes of ambushing them if they come to the ground or fall from the trees (as happens frequently among immature capped langurs).

SNAKES. The boas of the neotropics and the pythons of the Old World eat many species of small and medium-sized mammals. Boas and pythons are ambush hunters that eat infrequently and may sit for many weeks waiting for prey to cross their paths (Greene 1983). We know very little about the ecology of these large snakes; in fact, there has been no intensive field study of any of the largest snakes in Africa or Asia.

In the New World, boa constrictors have been reported preying on cebus (Chapman 1986) and squirrel monkeys. The African rock python, *Python sebae,* occurs in a wide variety of habitats. An adult reaches 6 meters in length and is capable of eating adult primates of all species except chimpanzees and gorillas. Only a few studies of their feeding habits have been conducted, and none has implicated primates as important food sources (Starin and Burghardt 1992). Instead, small forest antelope such as duikers (*Cephalophus* sp.) are the usual diet (Pitman 1974). In Abuko Nature Reserve in Gambia, E. Dawn Starin and Gordon Burghardt (1992) reported three instances of African pythons eating red colobus monkeys; it is not known whether the monkeys had been hunted or scavenged.

The large pythons reach their greatest diversity in Asia, where two very large and powerful species occur: the Indian python *(Python molurus)* and its geographic variants in South and Southeast Asia, and the reticulated python *(Python reticulatus)* in Southeast Asia. Although able to take primate prey, they are primarily terrestrial snakes when adult and do not often encounter arboreal primates. This fact, along with their slow metabolism and very infrequent feeding, suggests that they may not have much of an impact on wild primate populations.

HUMANS AND OTHER PRIMATES. Nonhuman primates are not only frequent prey items but also can be predators, although rarely on other primates. Butynski (1982) estimates that 80 percent of all observed mammalian predation events by primates have been by either baboons or chimpanzees. The best-known nonhuman primate that eats primates routinely is the chimpanzee, but it is not the only example. Olive baboons *(Papio anubis)* prey on red colobus occasionally, though these kills are sometimes pirated by chimpanzees (Morris and Goodall 1977).

In tropical forests in many parts of the world today, the principal predators on primates are humans armed with bows and arrows, blowguns, or firearms. Although human hunting pressure on primates has increased exponentially as human population density has exploded in the developing countries where most primates live, the human impact on nonhuman primates is not necessarily of recent origin. On the contrary, it is possible that in the Old World humans of one form or another have been hunting primates for food for thousands or even millions of years. *Homo erectus* was hunting game in African and Asian habitats several hundred thousand years ago; whether its diet included primates is unknown. Even the earliest hominids shared woodland and savanna habitats with primates similar to those hunted today by chimpanzees. Modern primate populations may have been exposed to human predation since antiquity. Hunting by humans may therefore be an evolutionary influence on the social system of primates, in addition to a modern proximate force driving many populations toward extinction.

Modern foraging peoples who share their forests with other primates also hunt them. The Aché of Paraguay are accomplished hunters (Hawkes et al. 1982; Hill and Hawkes 1983). They hunt mainly the larger neotropical monkeys, perhaps because the time and energy required to capture the smaller ones would not produce a nutritionally worthwhile payoff. *Cebus apella* in particular is a favored prey species of the Aché, and individuals are killed nearly every day of the year. Hunting success increases with size of the hunting party. Monkeys are pursued from the ground as they flee through the forest canopy; when they stop, they are shot with bow and arrow. The use of shotguns gives the Aché more meat per hour spent hunting for large mammals, but shotguns are not more effective in obtaining monkey meat than bows and arrows (0.02 kilo meat per man-hour for shotgun hunting versus 0.14 kilo per man-hour for bow and arrow). The Aché kill enough cebus, including entire groups, that they may be the limiting factor on cebus population size in their forest habitat.

Hudson (1991) points out that the effect of human hunting on prey populations is dependent on the degree of selectivity of that prey by the hunters. This selectivity can be based on the prey's behavior, which may render it a desirable target, or on the foraging return rate expected when prey other than only primates is sought (Hawkes et al. 1982). Small-bodied primates such as callitrichid monkeys are rarely targeted by human hunters, perhaps because they are simply too small to be worth hunting. The larger neotropical species, such as the 3-kilogram cebus or the 10-kilogram spider monkey or howler, are more avidly sought.

Selective predation by humans can also involve killing one age or sex more than others. For instance, among the Machiguenga and the Piro of southeastern Peru, all terrestrial mammals are killed by shotgun or bow and arrow in proportion to the age and sex structure of the populations. Primates, especially capuchins, are ambushed after being lured into killing range by imitations of infant cebus distress calls (Alvard and Kaplan 1991). Excluding young infants that are killed along with their mothers, 90 percent of primate kills are adults. Moreover, kills of primates differ from those of terrestrial mammals, in that bow hunting kills more females than expected and shotgun hunting kills more males (Alvard and Kaplan 1991). The Machiguenga have a very low success rate hunting monkeys with bows in spite of long pursuits. Consequently they rarely hunt capuchins, preferring to search for larger, more profitable prey. The Aché ignore capuchins even when hunting with shotguns.

The Waorani of the western Amazon Basin have the same preference for large-bodied monkeys as prey, and their use of shotguns has greatly increased their hunting efficiency and kill rates (Yost and Kelley 1983). The use of firearms changes the equation of hunting time and preferred prey because the odds of making a kill are higher with the modern technology. Choice of weapon may depend on local ecology and animal communities. The recent introduction of guns to the traditional arsenals of hunters has probably affected wild primate populations severely, compared to the effect of humans armed with bows and arrows.

Although neotropical foraging people are the best-known monkey hunters, hunter-gatherers in the Old World also eat primates. Bushmeat is eaten in many Central and West African countries; the advent of firearms poses an ever-increasing threat to the existence of forest primates, including great apes. Anadu et al. (1988) studied bushmeat consumption in Nigeria and found that red colobus monkeys were the primate meal preferred over all other species. Amman (1996) has documented the commercial trade in bushmeat by professional hunters.

Perhaps the most widely cited example of human predatory influence on wild primate populations is the hunting by people of the Mentawai Islands of two primates endemic to those islands. The Mentawais and associated islands off the coast of Sumatra harbor several primate species found nowhere else. Two of these species, *Presbytis potenziani* and *Simias concolor,* are reported to be monogamous; this monogamy has been linked to a recent history of risk of predation by the islands' human inhabitants. Ronald Tilson and Richard Tenaza (1977) hypothesized that monogamy might be a response to intense hunting pressure, in that small social units would be harder for hunters to locate than large social groups. Subsequent research, however, has shown that *P. potenziani* is not strictly monogamous; the role of predation in shaping the social system of this species is unresolved (Fuentes 1994).

I had taken Hugo up to show him the Peak and we were watching four red colobus monkeys that were evidently separated from their troop. Suddenly an adolescent male chimpanzee climbed cautiously up the tree next to the monkeys and moved slowly along a branch. Then he sat down. After a moment, three of the monkeys jumped away—quite calmly, it appeared. The fourth remained, his head turned toward the chimp. A second later another adolescent male chimp climbed out of the thick vegetation surrounding the tree, rushed along the branch on which the last monkey was sitting, and grabbed it. Instantly several other chimps climbed up into the tree and, screaming and barking in excitement, tore their victim into several pieces. It was all over within a minute from the time of capture.

—Jane Goodall, *In the Shadow of Man* (1971)

4. Chimpanzees as Predators

The morning of 3 September 1994 is heavily overcast, with the chance of a late–dry season rain shower. A party of male chimpanzees spend the predawn hours in the upper reaches of a small ravine called Hilali's Korongo, north of the main camp. Research assistant Bruno Herman and I accompany them. As the dawn chorus of tinkerbirds, bulbuls, and robin chats begins to ebb, eight silhouettes emerge from their night nests in the twilight and descend to the leaf litter below, where they sit like rocks. A charging display by the alpha male Freud wakes the party and they head slowly down the ravine, arriving shortly after 0700 in a clump of miombo overlooking Lake Tanganyika. Here they breakfast: the adult males Freud, Frodo, Wilkie, and Prof, plus the adolescent males Pax and Tubi. Two younger juvenile males, Kris and Sheldon, are traveling with the other males today. They are still in the limbo period of chimpanzee life between the security of a mother's company and the male-bonded life of an adult.

As the animals feed in scattered trees overlooking the water, the serenity of the scene is shattered by Prof's waa *alarm bark. The* waa *is given in response to a perceived threat from a snake, a strange chimpanzee, or—in the case of unhabituated chimpanzees—an approaching human. Prof leaps from his feeding tree and charges up the ridge, hair bristling. Neither Bruno nor I hear or see any source of potential danger, but the other chimpanzees respond immediately. Freud displays again, rushing past Prof while pulling clumps of brush from the ground. Suddenly the chimpanzees stop and are quiet; they sit clustered on the spine of the ridge and stare to the north. We sit behind them and listen for several minutes. Bruno offers the opinion that Prof heard a stranger's pant-hoot call coming from the northern Mitumba chimpanzee community. The Mitumbans have* been seen recently in Rutanga Valley, a half-kilometer north of where we sit. As if to confirm Bruno's suggestion, the males set off to the north in single file, breakfast forgotten, with the former alpha male Wilkie in the lead. They walk a short distance to the southern ridge of Linda Valley, then sit again and scan the valley silently. Their stillness and single-file formation suggest that they have heard a stranger's call and are now in territorial patrolling mode.

As we sit on Linda Ridge, we hear the high-pitched whistles and chirps of red colobus from across the valley. A large group of about 40 animals appear. They are in a thick stand of mbula (Parinari curatellifolia) trees growing on the thicketed north slope of Linda Valley. At 0800 pant-hoots erupt again from the Mitumba chimpanzees, this time clearly audible to all of us. Wilkie heads down a trail into Linda, followed by the rest of the party. The trail plummets through thick brush to the streambed, then leads up the other side toward the ridge separating Linda and Rutanga valleys. Halfway up the slope the party turns east, off the trail through dense thorn thickets of Smilax sp. This route takes them close to the colobus group, though Bruno and I are in whispered agreement that the males will not stop to undertake a hunt while patrolling. However, as the group passes just west of the colobus, Frodo races toward the mbula trees, his hair bristling. Freud and all the others follow. I scramble on hands and knees to keep up as the chimpanzees dash into the thickets that surround the trees where the colobus are located. It is now 0842 and we are in a drainage ravine of Linda Valley known as LK8.

The colobus have been caught entirely by surprise. Because they are in trees of modest height (12–15 meters) on a steep slope, the colobus in the tree crowns are nearly at eye level with chimpanzees higher up the slope. Crawling toward the trees from the thickets, I see that one side of a mbula crown is thick with colobus. The gap to the next tree is too great to leap across, and more and more colobus are crowding those already at the edge of the crown. At 0848 Wilkie races up this tree and grabs an infant left unprotected by its mother in the chaos; he bites through the skull and takes the tiny carcass back to the ground. Frodo catches a somewhat larger juvenile and disappears into the thickets. Meanwhile, Freud appears with a juvenile who apparently fell from the tree.

A moment later it begins to rain colobus monkeys. The colobus bunched in the tree crown above have reached a critical mass. In twos and threes they begin to fall and leap from the tree. They thump into the *leaf litter all around me; a juvenile female lands at my feet. She begins to scramble awkwardly toward the tree trunk but the young male Kris, who has remained on the ground, leaps on her and begins to bite her nape and back. Tubi races to the scene and chases Kris away. Freud appears and, uncontested, appropriates the now-immobile female colobus from Tubi. Freud stands bipedally for a moment, holding his own catch by the throat in one fist and his stolen prize in the other. At 0851, after nine frantic minutes, the hunt is over. Prof arrives on the scene carrying his own young colobus. The final tally for the day is five colobus killed—three juveniles and two infants. Nearly half of the female colobus in the group have thereby lost their reproductive output for this year.*

The postkill scene is unusually orderly. Each chimpanzee sits with his kill within a circle about 10 meters in diameter, some on the ground and some on low tree limbs overhead. Frodo has disappeared and is not seen again, though I later find the remains of what was probably his kill some 15 meters away in dense thicket. The sharing of meat ensues. Freud has two carcasses and about 8 kilograms of meat. Tubi approaches him and begs for meat; he is given a morsel, then is allowed to take almost the entire larger carcass—the one Freud had stolen from him in the first place. Tubi then shares this returned carcass with Kris, the original captor, and with Prof, who has finished his own meat and comes begging for more. Pax receives some meat from Prof, while Freud shares his remaining carcass with Kris, Prof, and Tubi. Wilkie does not share and Sheldon departs the scene, having received no meat. All the hunters eat leaves from nearby undergrowth with their meat. More than three and a half hours later, the hunters are still lying on the same spot, having eaten, napped, and then eaten more. At 1227 Freud departs, followed by the remaining males. They leave bone fragments scattered in the dry leaves and most of one carcass uneaten. The party moves off to a patch of mabungo makubwa *fruit* (Saba florida), *their only food other than meat since early morning.*

Hunting Patterns

The hunt depicted above is atypical in several ways. First, it involved only male chimpanzees. The colobus carcasses were not eaten entirely—normally every scrap of bone, meat, and hair is consumed—and the prey were captured while fleeing in chaos rather than while counterattacking. The description does, however, reflect the efficiency of chimpanzees as predators and their potential for exacting, within a few minutes, a tremendous toll on colobus groups.

Their desire for meat notwithstanding, chimpanzees are primarily frugivores; ripe fruit accounts for up to 70 percent of the annual diet (Wrangham 1977; Goodall 1986). They also consume substantial amounts of leaves, flowers, seeds, and insects, depending on availability. Although they feed on more than 30 species of mammal (Uehara 1997), chimpanzees may consume overall more invertebrate than vertebrate biomass through termite and ant-fishing. (The weight of insects and the amount of nutrients obtained is hard to estimate; McGrew 1992.) The eating of meat was one of the dramatic discoveries reported first by Goodall (1968) in the early years of her field study. Long-standing assumptions about the exclusively herbivorous nature of the apes fell, and meat-eating—along with tool manufacture and use—confirmed the chimpanzee as the model of choice for human evolutionary studies.

Compared to the other three species of great apes—bonobos of Congo rain forests, gorillas of equatorial African forests, and orangutans of Indonesian rain forests—chimpanzees are notably carnivorous. Even after Goodall (1968) reported it, the extent of meat-eating was underestimated until quite recently. When Geza Teleki undertook the first systematic study of chimpanzee predatory behavior in 1968–1970, he recorded only 4 red colobus (19 percent) in a total of 21 mammals killed (Teleki 1973). The majority of kills in Teleki's study (12 of 21) were olive baboons. Teleki analyzed data from the 1960s and found that only 14 of 56 total kills over 11 years (25 percent) were red colobus, while 21 of the 56 (38 percent) were baboons. These figures probably reflect the influence of provisioning with bananas rather than a natural tendency of chimpanzees to prefer baboons as prey over other animals.

In the 1970s, Busse's (1977, 1978) study of hunting of red colobus by Gombe chimpanzees showed that the chimpanzees did not hunt during every encounter with their prey and that their success rate was less than 50 percent (in 31 of 64 hunts, one or more colobus were caught). Meanwhile, Japanese researchers in the Mahale Mountains, 100 kilometers away, were also finding that chimpanzees were avid predators of mammals. Chimpanzees hunted the same animals as at Gombe, plus blue duiker antelopes (*Cephalophus monticola*). Red colobus were an important prey, but in some years the duiker, which do not occur at Gombe, constituted more than half of the meat diet. Nishida, Uehara, Takahata, and others studied predatory behavior at Mahale and showed that it was more seasonal in occurrence than at Gombe (Takahata et al. 1984).

Little more was learned about chimpanzee hunting until the Swiss zoologists Christophe and Hedwige Boesch reported predation by chimpanzees in Taï National Park, 4,000 kilometers to the west in Ivory Coast (see Table 4.1). The Boesches, working in a lowland rain-forest habitat, reported that their hunters showed a different pattern of predation than had been observed at either Gombe or Mahale. At Gombe, Busse had claimed that hunting, though often a communal activity, was essentially noncooperative. That is, adding more hunters to the attack on colobus monkeys did not enhance the success rate. At Taï, the Boesches reported elaborate coordination during hunts of red colobus, to the extent that the hunters appeared to take roles, increasing the odds that a colobus would be caught (Boesch and Boesch 1989). Females were more involved in the hunt than at Gombe, and after a kill females more actively participated in the distribution of meat. In addition, the Boesches stated that Taï chimpanzees actively searched for prey, whereas researchers at Gombe and Mahale saw only opportunistic hunting in the course of foraging for plant foods. The reports by the Boesches of planned, well-coordinated hunting efforts taking place in a rain-forest environment were among the most provocative findings about chimpanzee behavior to emerge during the 1980s.

The Variety of Prey

Red colobus constitute the majority of prey at all long-term chimpanzee study sites. At Gombe they are numerically the overwhelming favorite (Figure 4.1). From 1990 through 1994, nearly 85 percent of all mammalian prey items were red colobus. Other prey included bushbuck antelope, bushpigs, baboons, and red-tailed and blue monkeys.

Table 4.1. *Outline of hunting patterns of Gombe chimpanzees (1990–1995), compared with those of other chimpanzees at other study sites. (Adapted in part from Stanford 1996.)*

PATTERN	GOMBE	MAHALE[a]	TAÏ[b]
Major prey species (as a percentage of all mammalian prey)	Red colobus (84.5%)	Red colobus (53%)	Red colobus (78%)
Mean hunting success rate (red colobus)	54.5%	?	57.3%
Mean length of successful hunt (minutes)	36.9	18.2	15.2
Correlation between number of hunters and likelihood of success	Yes	?	Yes
Percentage of kills by males	89%	79%	71%
Seasonal hunting peak	Yes	Yes	Yes
Percentage of kills in peak season	34.6% (Aug.–Sept.)	60.0% (Oct.–Nov.)	? (Sept.)
Hunting peak	Middle and late dry season	Early wet season	Wet season
Prey age classes for red colobus	Adults 10.6% (males 1.8%, females 8.8%, immatures 89.4%	Adults 30%, immatures 70%	Adults 47%, immatures 53%

a. from Takahata et al. 1984; Nishida et al. 1983; Uehara et al. 1992.
b. from Boesch and Boesch 1989; Boesch 1994c.

Of these, bushpig and bushbuck were the main alternatives to red colobus (8.1 percent and 5.3 percent of the prey items, respectively).

Little is known about the ecology of either bushpig or bushbuck, but field observations suggest that both are seasonal breeders. Bushpiglets appear mainly from August through March, with a pronounced peak in September, suggesting that this is the peak of the bushpig birth season (Gombe Stream Research Center, unpublished data). The timing of bushbuck sightings is more constricted; most of the sightings between 1990 and 1994 were in June (Appendix 5). No chimpanzee would tackle

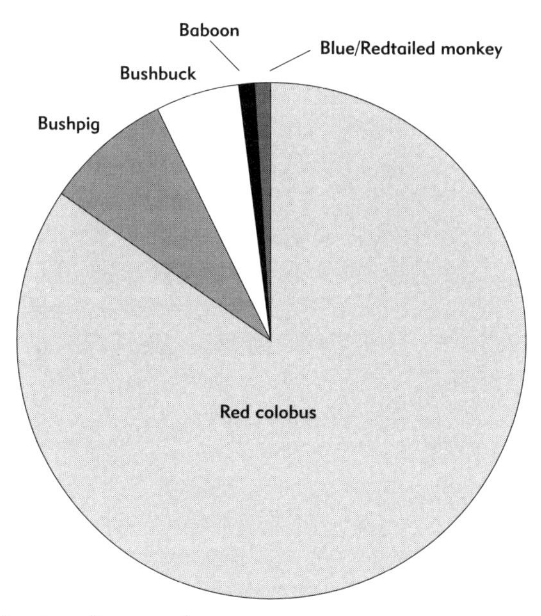

FIGURE 4.1. Portion of mammalian prey species in the Gombe chimpanzee diet, 1990–1994.

a pig or bushbuck that is more than half grown; the pigs in particular are belligerent and can use their tusks dangerously. Instead, fawns and piglets are taken when encountered in dense thicket. The hunter grabs and makes off with a squealing baby while the parents attack or chase the hunters. The occurrence of most incidents in thickets, the snatch-and-run tactic used by the chimpanzees, and the chaotic moments after these encounters, all make observations of hunts of these species difficult. In some cases the chimpanzees startle adult pigs, then appear to search the leaf litter for piglets. In most cases, however, the hunts involve little planning and bear little resemblance to hunts of red colobus.

Goodall (1986), Wrangham and Bergmann-Riss (1990), and Uehara (1997) describe the variety of prey animals eaten by East African chimpanzees. The vertebrate animals eaten total at least 35 species. I have seen foraging chimpanzees chase monitor lizards *(Varanus niloticus)*, frogs (*Rana* sp), and birds encountered on the ground. After happening upon a litter of African civets (*Civettictis civetta*) in a thicket, juvenile chimpanzees used them as playthings rather than as food (perhaps because of the musky odor for which civets are well known). When capturing primates other than red colobus, chimpanzees often use a flat-out chase rather than a coordinated group hunt. In October 1994 Frodo casually approached an infant baboon who was foraging near its mother and a large adult male. The male baboon sensed that Frodo's approach was not innocent; the male grabbed the baby and Frodo gave chase. He chased the male baboon, who was more than half Frodo's size, through tree crowns for nearly 100 meters with the baby clinging to the male's chest, until Frodo gave up.

Although baboon kills are rare today compared to earlier eras at Gombe, they do occur and chimpanzees definitely see them as potential sources of meat. Similarly, the only successful hunt of a guenon that I observed was a mother-infant blue-monkey pair that Frodo saw while foraging on the forest floor. Frodo was more than 75 meters away from the blue monkey in an open patch of forest; he immediately set off at high speed, apparently able to see that the blue monkey was carrying a newborn baby. He chased her through several thickets, emerging finally with the baby monkey, which weighed no more than 100 grams.

These rapid ambushes or pursuit hunts occur very rarely with red colobus, perhaps because they spend so little time on the ground or away from the protection of their group.

Chimpanzees spend considerable time attempting to capture red colobus monkeys. The mean average length of hunts that I observed between 1991 and 1995 was 28 minutes (17 minutes for failed hunts, 37 minutes for successful hunts). This figure is higher than previous estimates for Gombe (Stanford et al. 1994a) because of several extremely long hunts in the sample. I observed five hunts longer than an hour and one longer than an hour and a half. The amount of time chimpanzees spend consuming the body tissues of other vertebrate animals (hereafter called meat) has been estimated at between 1 and 5 percent (Teleki 1973, 1981; Goodall 1986) of their feeding time.

Chimpanzee omnivory is deceptive, as an analysis of the amount of meat consumed annually reveals (Figure 4.2). Based on estimates of the body weight of red colobus, the Kasakela community consumed more than 500 kilograms in some years. Combined with the biomass of other prey species eaten, the total quantity of meat consumed by the 45-member community approaches 700 kilograms. This figure is similar to an estimate of 441 kilograms per year made by Wrangham and Bergmann-Riss (1990), based on 1970s data from Gombe. The quantity

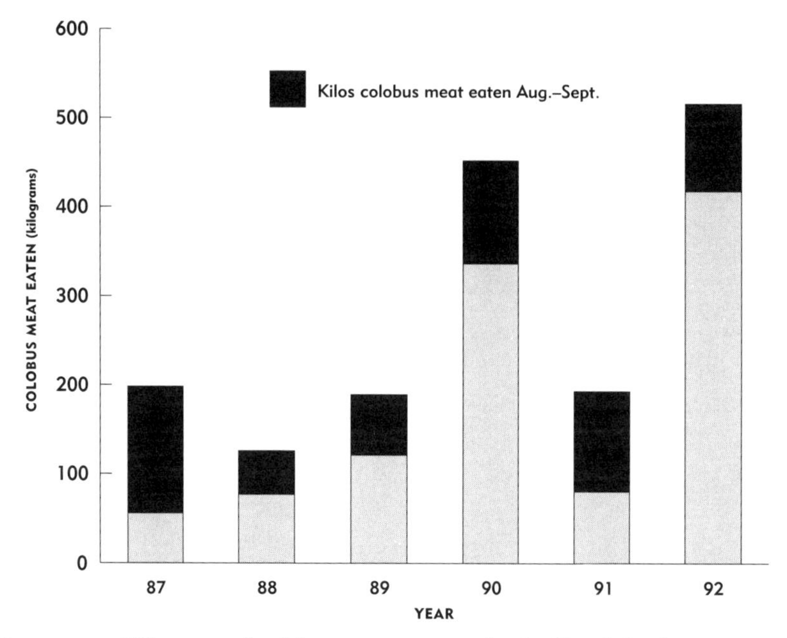

FIGURE 4.2. Kilograms of colobus eaten per year by the Kasakela chimpanzee community, 1987–1992, with kilos eaten during the August–September peak in horizontal hatching. Reprinted with permission of *American Anthropologist*.

of meat apparently has been consistent for the past two decades and is not the result of a change in meat-eating patterns in recent years, in spite of changes in the number of male hunters (Appendix 6).

Moreover, the division of meat among the members of the community is far from equal. Hunters themselves consume far more meat than females do, and peripheral females and immature chimpanzees of both sexes are often left out of meat-eating episodes altogether. I estimated that on a per-capita basis each of the adult male hunters consumed about 0.5 kilogram of meat per week during the peak periods of meat-eating (Stanford 1996). This figure fits well within the lower end of the spectrum of caloric intake of meat protein (Bailey and Peacock 1988 for Efe pygmies of Congo). Unlike human hunters who sometimes catch large mammals, all of the meat eaten by chimpanzees comes from small animals.

Seasonality

Hunting by Gombe chimpanzees is seasonal, a fact that has important implications for understanding chimpanzee hunting patterns and their effects on their prey populations. If chimpanzees hunt more at one time of year than at others, either the physical or the social environment must generate cues that influence the behavior. Or peaks may be related to the dynamics of forest ecology. For instance, if the timing of the hunting peak at Gombe were related to a birth peak in one or more of the prey species, then chimpanzees might be responding to independent ecological patterns by exploiting an ephemeral source of easily obtained meat. Hunting might be timed to coincide with periods of rainfall or fruit abundance; thus predator and prey might encounter each other more frequently in those periods.

None of the early work at Gombe (Teleki 1973; Wrangham 1975) suggested that hunting was seasonal, probably because these studies involved small sample sizes. On the other hand, Mahale researchers had shown that hunting was sharply seasonal, the peak occurring in the early-rainy-season months of October and November (Takahata et al. 1984). The Taï chimpanzees also hunt seasonally, with the peak at the height of the rainy season (September) in Ivory Coast (Boesch, personal communication).

In 1991 I analyzed Gombe data from the 1980s and found a number of previously unknown hunting patterns. The Kasakela chimpanzees hunt most frequently in February and from July to October (after correcting for observer hours; Stanford et al. 1994a; Figure 4.3); Appendixes 7 and 8). This effect is due to a nonrandom clumping of hunting activity during these months, peaking in August and September, which are the height of the dry season in western Tanzania. An analysis of the rate of hunts per 100 observer hours in the forest produced the same result. In all, 35 percent of all hunts recorded took place in August and September from 1982 to 1992 (Stanford et al. 1994b); between 1990 and 1995, 30 percent of hunts occurred during these months. Hunting success is also higher in the dry season than in the rainy months. The reason may be the greater visibility due to lack of foliage, which enables the hunters to select their targets and see what the colobus are doing.

Success Rates

Just as hunting frequency shows month-to-month variation, the success rate of hunting varies. At Gombe (no data are available for the other chimpanzee study sites) the monthly mean success rate for hunting parties varies between a low of 35.7 percent (Figure 4.4, December)

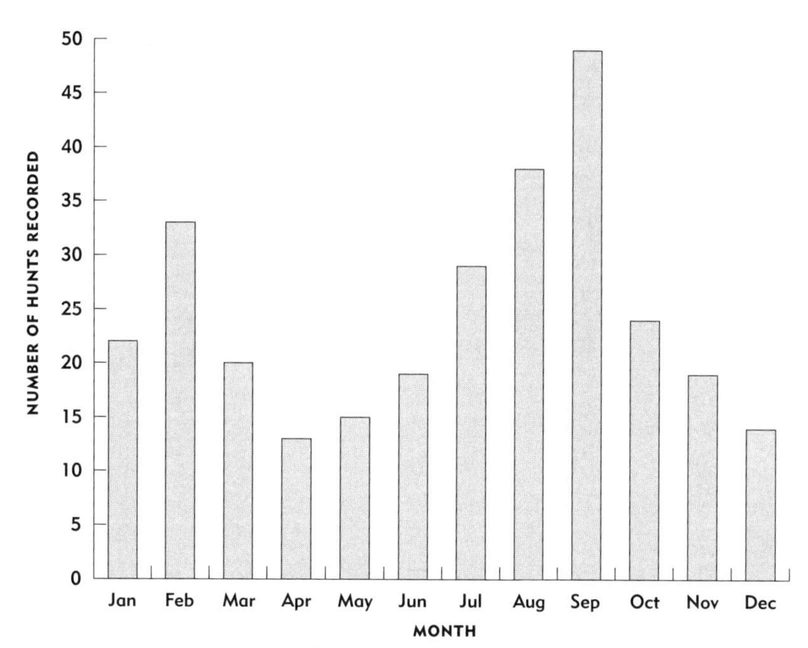

FIGURE 4.3. Hunts of red colobus observed per month at Gombe, 1990–1994, all groups.

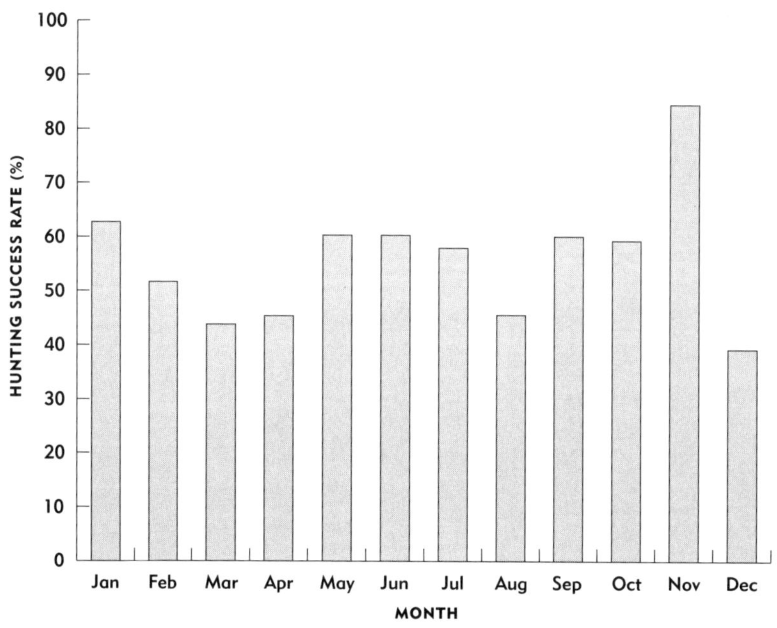

FIGURE 4.4. Monthly variation in the percentage of hunts at Gombe in which at least one colobus was caught, 1990–1995.

and a high of 84.2 percent (November). There is no correlation with any aspect of party size or other social parameters that differ between November and December.

An earlier analysis had shown a peak in hunting success in the dry season that was not statistically significant (Stanford 1994a). The success rate was significantly positively correlated with the number of males in the hunt (Figure 4.5) and with overall party size (Figure 4.6). Why more hunters should lead to greater hunting success seems clear; more attackers mean more opportunities for at least one to make a kill.

Why hunting success should be tied to overall party size is not as obvious. Larger parties are very likely to have more males and therefore more hunters, but the independent influence of party size (Stanford et al. 1994b) suggests other benefits. When a chimpanzee foraging party attacks a red colobus group, some members of the party climb the tree holding the monkeys while others remain on the ground. These terrestrial bystanders are also involved in the hunt; they follow the course of the action from beneath, and some enter the trees upon seeing a colobus about to escape. For this reason I questioned Boesch's consideration of hunters as only those members of the party who are pursuing colobus (Stanford 1996). The significance of overall party size is borne out by the tendency of larger parties to be more successful at making kills regardless of the number of hunters present. Another effect of party size is that with more chimpanzees present, important events are likely to occur out of view of the observer. In the chaos of the hunt, with foliage obscuring one's view, chimpanzees who are bystanders at the beginning of the hunt often become leading players later on. This can happen either because a colobus falls from a tree onto the forest floor near a bystander or because the bystanders become active participants.

My earlier studies showed that the microhabitat in which in the hunt occurred could be a major influence on hunting success (Stanford 1994b), confirming findings by Wrangham (1975) and Boesch (1994c). Microhabitat and terrain influence the course of the hunt in the following ways. When chimpanzee parties encounter colobus in areas of forest in which the canopy is tall (more than 20 meters at Gombe) and continuous, they must climb higher to attack their prey. The colobus have more escape routes in continuous canopy, and have more time to organize themselves to defend against the attack. In low

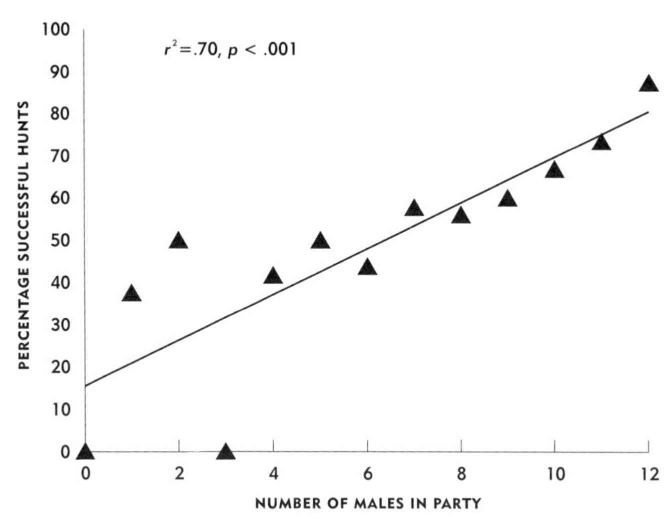

FIGURE 4.5. Relationship between the number of male chimpanzees in the hunting party at Gombe and their rate of success, all colobus groups.

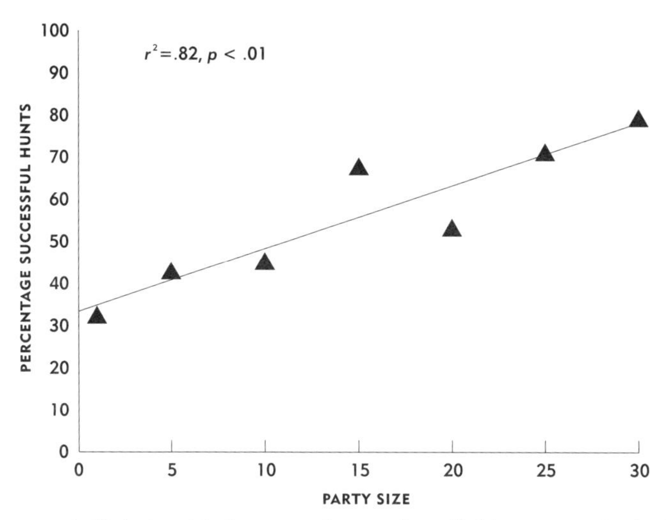

FIGURE 4.6. Relationship between the number of chimpanzees in the hunting party at Gombe and their rate of success, all colobus groups.

broken canopy, the chimpanzees reach their quarry quickly. The monkeys cannot move easily to another tree, leaving themselves exposed before the male colobus can mount a counteroffensive. Hunts in these situations are more often successful than in continuous canopy, and kills are made much more quickly (Stanford 1995).

For example, on 3 July 1994 a party of 18 chimpanzees, including 9 adult and adolescent males, hunted the AK colobus group for more than an hour in the canopy over the Kasakela streambed. The forest there is quite tall and the riverine trees along the banks of the dry streambed make a long, continuous travel path for colobus under attack. The AK group moved up and down the stream; its males defended the females and immatures who stayed above and behind in the face of chimpanzee attacks. Finally, after 65 minutes and several lulls, during which the chimpanzees rested and seemed to size up their prospects for a kill, Frodo managed to scatter the five counterattacking males of AK group long enough to catch a juvenile colobus (see photo insert). Hunts with a similar complement of hunters often end within a few minutes when they occur in open woodland.

On 7 October 1992, the 25 members of the colobus J group were feeding in the broken canopy of a hillside in Kakombe Valley known as Dung Hill. From dawn until late morning they moved slowly and uneventfully across the slope and into a ravine known as KK6. The colobus were relaxed as they fed on foliage and young fruits. Beginning at about 0900, the pant-hoots of one or more chimpanzee foraging parties could be heard from farther down the valley. The 5 male colobus gave occasional high-pitched alarm calls—but the chimpanzee calls came little closer. Then, at about 1100, chimpanzee pant-hoots erupted from both north and south of the colobus group. For several minutes two chimpanzee parties called, then the calls converged.

Minutes later, Beethoven arrived with several of the adult females and their offspring. They were followed by Tanzanian observers Msafiri Katoto, Bruno Herman, plus Charlotte Uhlenbroek, a British student. In a few moments, all of the community's 12 adult and adolescent males and many females and juveniles arrived—33 animals in all. The hunt began with Frodo climbing a tall emergent tree in which some of the colobus group were clustered. For the next 20 minutes, the trees shook and the foliage crashed with the sounds of frightened colobus and chimpanzees. With so many hunters and only 5 colobus males defending, Frodo and the other males managed to scatter some of the male colobus, whereupon the rest of the group fled and became easy prey. A juvenile colobus attempted to escape by leaping onto a branch where Atlas sat. The chimpanzee captured the monkey and killed it with a bite to the base of the skull (see plates 15–16). A few meters away,

Beethoven caught an infant colobus and shared the carcass with Gremlin. After about an hour, some chimpanzees were still hunting, while others who had captured colobus sat on the ground over a 50-meter circle eating and sharing meat. After nearly 90 minutes of hunting, six colobus had been killed and eaten, plus a seventh whose carcass was being carried by its mother. Four hours later, the chimpanzees were still finishing their feast.

A HUNT SUCH AS THIS does not occur often at Gombe; indeed, this was only the second kill of seven colobus observed in more than three decades of research. But multiple kills of two or more colobus happen frequently, 21 times in 1990 alone, illustrating the powerful influence of chimpanzees as predators on the populations of prey animals within their hunting range (Figure 4.7). Quite often, chimpanzees kill several members of a colobus group in the same hunt. A chimpanzee party that attacks a large colobus group repeatedly can decimate it by virtue of the number of prey taken in predatory episodes. Although one would expect the odds of making a multiple kill to increase in direct relation to the number of male hunters present in the party, this is not actually the case. There is not a significant correlation between the number of

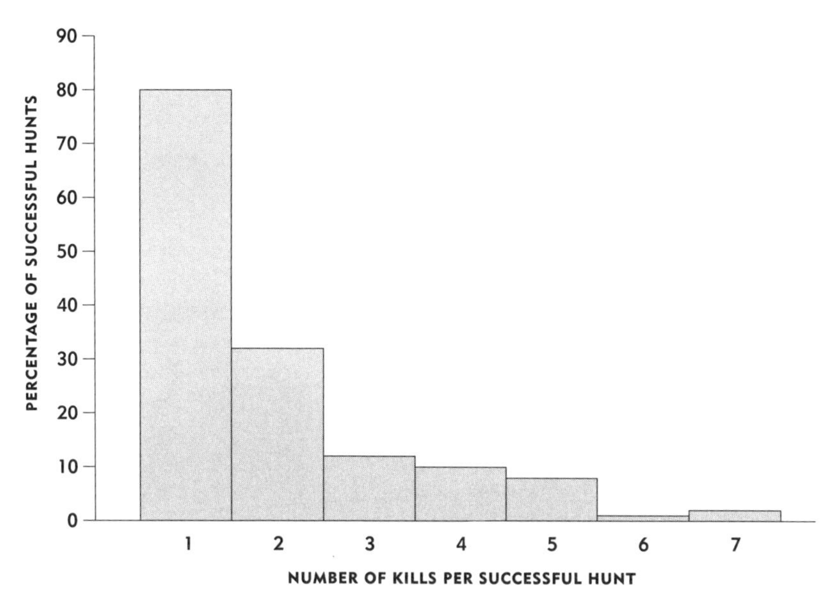

FIGURE 4.7. The number of kills per successful hunt at Gombe, 1990–1995.

kills and the number of hunters ($r^2 = .097$, $p = .06$), and a stronger positive correlation exists between the overall hunting party size and the number of kills ($r^2 = .247$, $p < .001$). Hunts in which multiple kills are made last significantly longer than single-kill hunts (37 minutes versus 17 minutes, $t = 6.87$, $p < .001$).

Age and Sex of Colobus Prey

Some chimpanzee populations hunt mainly adult red colobus, and others kill primarily juveniles or infants. At Gombe the pattern since records have been kept is that the majority of kills of red colobus are of immature animals. All analyses have shown that the percentage of kills that are young animals is significantly greater than their representation in the Gombe red colobus population (Figure 4.8). Goodall (1986) showed that in the 1970s three-fourths of all kills were of immatures, and my group's analysis for the 1980's showed a nearly identical pattern. Between 1991 and 1995, I observed 57 red colobus kills (of a total of 76) in which the victim could be reliably categorized as to age or sex. Of these, only 5 females and 1 male (10.6 percent) were adult animals, and 2 were subadults of indeterminate sex (3.5 percent). Thus 85.9 percent of these kills were of immature animals, significantly greater than

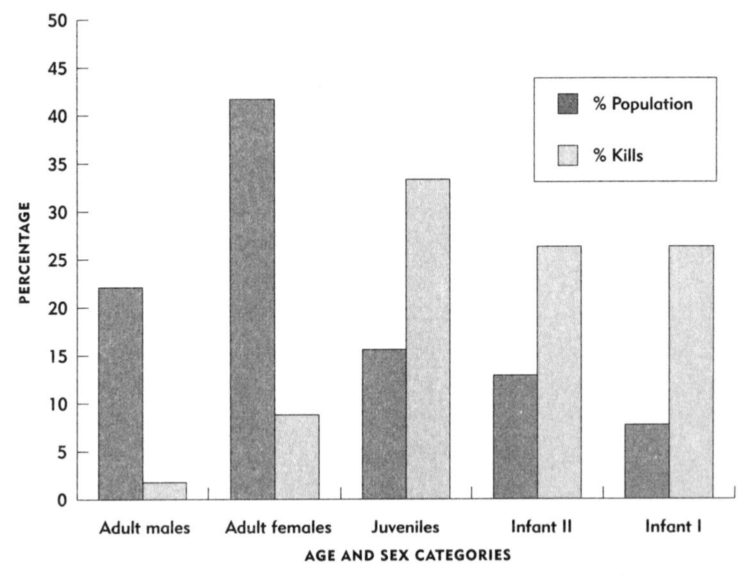

FIGURE 4.8. The age and sex categories of the red colobus population at Gombe compared to the age and sex categories of red colobus killed by chimpanzees.

the 42 percent immatures of all red colobus censused at Gombe (one-tailed t-test; $t = 8.77$, $p < .001$) and dramatically greater than the adult-to-immature ratio of the five colobus study groups, which were only 25 percent immatures.

At Mahale the ratio of adult colobus prey to immatures is similar to that at Gombe (Takahata et al. 1984). These figures contrast sharply with Taï, however, where nearly half (47 percent) of all kills are adults, including many adult males. Although detailed information on the Taï red colobus population does not exist, infants presumably make up at least 40 percent of the population. The Boesches (1989) reported a Taï chimpanzee preference for adult colobus; they capture adult males when hunting because they are mobbed by these males, and when capturing an infant they sometimes also take the mother. The Taï chimpanzees apparently have little choice of size of acive prey. The large percentage of adults occurs not because of an aversion to immatures or a preference for adults, but instead because adults are captured in roughly equal proportion to their number in the population.

There are two likely reasons for the preponderance of immature prey at Gombe. First, chimpanzees there may have more difficulty or incur more risk capturing adults than at Taï. Alternatively, they may have little difficulty capturing adults but nevertheless prefer to eat immatures. Most observers of hunting at Gombe have advocated the former explanation. Gombe red colobus are so aggressive in defending their offspring against chimpanzees that hunters are sometimes injured by bites from the males, perhaps leading them to turn to more easily caught infants and juveniles. Though there is some truth in this observation—juveniles in particular are easily caught relative to adults and to infants—evidence supports the latter explanation. First, the hunters who seem (to human observers) to be the most skilled catch among the highest percentages of immature colobus, suggesting that they do so through choice. Frodo in particular is able to capture red colobus of any size without regard to the counterattacks of the male colobus, but consistently chooses newborn infants clinging to the abdomens of their mothers. Second, hunters sometimes catch and kill adult red colobus, only to quickly discard them and continue to pursue immatures. Boesch (1994c) regarded this behavior as puzzling because of the much greater amount of meat available from an adult, but it is logical if the hunter's preferred target is an immature.

This preference for young colobus could be linked to a number of factors. Babies may be easier to monopolize for hunters who do not intend to share their kill. They may simply taste better to the hunters and to the potential meat recipients, making them more valued as social currency. If Gombe chimpanzees take mainly immature colobus because they are young and tender, they are unlikely to be capturing them for solely nutritional reasons. Instead, immature colobus may make better gifts of meat to be shared with others for social and political benefits.

Do Chimpanzees Use Search Images?

In the hunt of 3 September 1994 described at the beginning of this chapter, the chimpanzees sat for many minutes on a ridge looking straight at the territory of the northern enemy. Across the valley and east of them was a large and noisy group of red colobus. In the ensuing minutes the party of hunters crossed the valley, turning to the east away from the trail, which led due north. Since the party was able to hear the colobus at a great distance and then seemed to orient its direction of travel toward the colobus group, can it be inferred that the chimpanzees turned in that direction in order to hunt the colobus group?

This kind of intentionality may occur occasionally, but it is certainly not the norm. Chimpanzees possess a sophisticated mental map of their landscape and appear to know the location of important fruit trees within their territories. They may even know the location of thousands of individual trees—at least their general location—and orient to them while traveling between favored spots. While the ranging behavior of chimpanzees is well documented, their microranging is known only though the impressions of researchers. Perhaps parties zigzag between trees as they travel because they know exactly where they are headed. Alternatively, they may be encountering desired foods randomly as they travel between known food locations.

Previous researchers have determined whether group travel is in the direction of some resource such as water or food more often than random chance would predict, then watched to see if that is where the group is headed (Sigg and Stolba 1981). Neither at Gombe (Wrangham 1975; Busse 1977) nor at Mahale (Nishida et al. 1983) have investigators thought that chimpanzees were other than opportunistic in their hunting pattern.

Boesch and Boesch (1989) used impressions of hunting detours and observations of searching behavior to argue that the Taï chimpanzees sometimes (in about one-third of their hunts) formed search images and deliberately tried to encounter red colobus groups. Documenting this thesis should be straighforward; more often than would be predicted by chance, do lines of travel change toward the prey upon detecting the presence of a colobus group at a distance?

At Gombe this is rarely the case. During 60 encounters between colobus and chimpanzees in which I was traveling with the chimpanzees before the encounter, the direction of travel deviated toward distantly heard colobus calls in only four cases. It is much more common for chimpanzees to pass directly beneath a colobus group, stop and scan the canopy, then either begin a hunt or not. It is also usual for a chimpanzee party to hear a colobus group vocalizing 10 or 20 meters away from the trail and then leave it in order to hunt the group.

Even if search images are rare, chimpanzee parties may travel to parts of their range where colobus groups are commonly found with the expectation that they will meet colobus while there or en route. For instance, in 1994 the Kasakela chimpanzees traveled to KK5 on four consecutive days to feed on a large crop of *mgege* fruit (*Syzigium guineense*). They met the same two colobus groups there each day and hunted them twice, once successfully. Although by the fourth day the fruit crop was much diminished, the chimpanzees returned to the same trees and picked off the few remaining fruits. It is impossible to know whether they were motivated by the memory of meeting colobus in those trees previously, simply came to scavenge the last of the mgege, or perhaps some combination of the two factors.

Binges

From the early days of Gombe research, Goodall noted that the chimpanzees periodically had hunting "crazes," during which many colobus or baboons would be caught. She noted that after returning from patrols to the borders of its territory, a party would sometimes hunt as though to vent the tension from hours of remaining silently vigilant (Goodall 1968). Hunting might also recur after a lapse of two weeks or more, or occur only sporadically for a number of months, whereupon a party would make a kill and then begin to hunt on a nearly daily basis. Earlier observers (Teleki 1973, Wrangham 1975) did not find any seasonal pattern in this behavior. I labeled these intense hunting periods "binges" and was eager to know why they occurred. Obviously such binges could decimate a colobus group if the chimpanzees continued to hunt that same group every day. As I developed a framework to test the purpose of hunting binges, several hypotheses seemed reasonable:

1. Hunting binges as such do not exist (the null hypothesis). Statistical tests might show an apparent clumped distribution of hunts. Such a distribution, however, could be an artifact of small samples of the observer's time in the field coincidental with observed hunting and the lack of recording of nonhunting periods.
2. Hunting binges are actually seasonal hunting patterns that are followed by seasons of little hunting.
3. Binges are caused by the presence of specific individual hunters who act as catalysts. When they are in the party, they hunt and others join in.
4. Binges happen because both chimpanzees and colobus are using the same food resource day after day in a particular place, so hunts occur frequently in that place.
5. Binges are linked to the birth seasons of prey species, including red colobus.
6. Binges occur, as reported by Goodall, mainly during periods when patrolling is frequent.

Before searching for explanations, it was necessary to define a binge. I defined it as any time period in which more than three hunts, successful or unsuccessful, occurred in a seven-day period. From 1990 to 1995, there were 23 binges, 15 of them between 1990 and 1992 (Appendix 9). In order to understand a pattern of hunting that included the tendency to binge, I reread the Gombe records back through the early 1980s. I found that binges occurred quite often during the peak hunting months of July to October, as we had already identified, but they could occur in any month. The pattern of hunting was not only significantly seasonal, but also clumped outside of seasons. Hunts were significantly more likely to occur on days after hunting than on days after nonhunting (one-tailed t-test; $t = 8.50$, $p < .01$). They were also more likely to occur within two days of a previous hunt ($t = 4.33$, $p < .05$).

This pattern was obvious when I was in the field with the animals. After weeks of infrequent hunting the chimpanzees would begin to hunt almost daily. I would then switch my attention from the colobus to the chimpanzees for the duration of their hunting spree. I did not find binges to be closely linked to male patrols to the territorial perimeter. Hunts during even the most intense binges took place across the entire home range of the community and were not limited to one area, or even to the same north or south direction of the core area. Instead, some unknown factor triggered the hunting binges of the Kasakela, and the tendency to hunt colobus when they were encountered continued for days or weeks.

The most intense hunting binge occurred in 1990, shortly before I began to study hunting. It began on 7 July and continued, with the chimpanzees hunting every other day or so, until 19 September—a total of 74 days in which 38 hunts were observed and at least 76 colobus were killed (Figure 4.9).

Many more unseen hunts may have occurred during this time period, since the fission-fusion society of chimpanzees precludes that the entire community can be observed continuously. If hunts occurred at the same rate when the chimpanzees were not being followed, then the total kills during this binge was more than 100. The chimpanzees

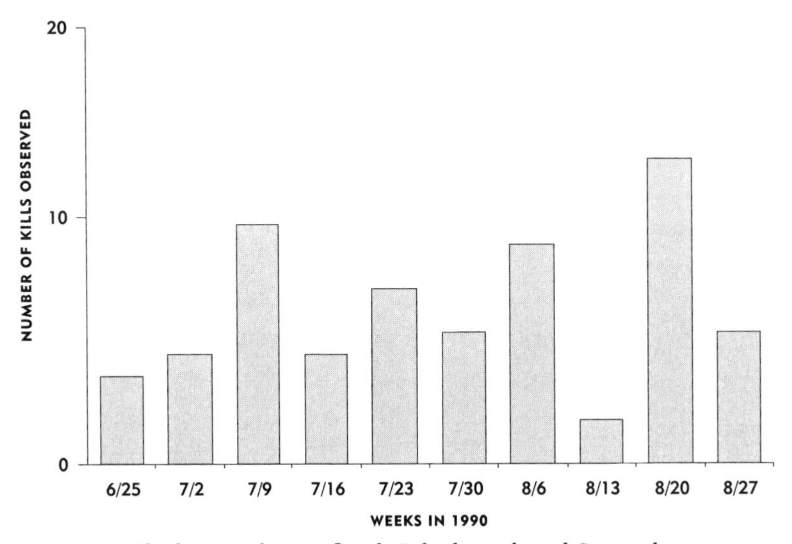

FIGURE 4.9. The hunting binge of early July through mid-September 1990, during which 76 colobus kills were observed in 74 days and 38 total hunts.

had also binged from 13 June to 25 June during which time five hunts were undertaken, resulting in four kills. So from mid-June through mid-September a minimum of 80 red colobus, and perhaps many more, were killed by the chimpanzees.

If I wanted to understand the reasons for binges, and perhaps by extension the reasons for hunting, looking for factors present in the community during this particular binge seemed an appropriate strategy. Some remarkable aspects of Kasakela chimpanzee grouping patterns were apparent. First, the mean party size from mid-June through mid-September 1990 was the largest recorded for any similar period in the past two decades: about 20 chimpanzees on average traveling together daily. Second, the average number of adult and adolescent males in parties was also higher than in any other two-month period on record at Gombe; all 12 males were almost constantly together. Finally, the number of females who were cycling, and therefore had large pink sexual swellings, was greater than at any other time recorded. On some days as many as 7 swollen females traveled in large parties accompanied by all the males. One or more of these three factors—party size, number of males, and presence and number of swollen females—seemed likely to account for the tendency to hunt during these periods.

Individual Variation in Hunting

Just as chimpanzees show extraordinary individual differences in personality, aggressiveness and political influence, so also do they vary in hunting performance. Some of Gombe's most politically influential male chimpanzees, such as Goblin, have also been skilled hunters. Other top-ranking males, such as Wilkie, were less eager and clearly afraid of male red colobus counterattacks. Hunters vary widely in their eagerness to hunt, their willingness to tangle with a trio of angry male colobus, and individual tactics of hunting. In some cases the presence of one powerful hunter tips the balance toward the chimpanzees in a hunt.

On 26 September 1991 I was following Frodo as he strode up the main trail in Kakombe Valley toward the waterfall. He was alone and no other chimpanzee call had been heard in the past half-hour. As we approached the waterfall, pant-hoots erupted from above the cliff, audible over the sound of the cascading water. Frodo instantly turned his head upward and scanned the treetops. This seemed odd since the calls were definitely coming from some distance away. Then Frodo turned back in the direction of the calls and rushed off, crossing the stream and climbing a steep trail next to the falls to reach Upper Kakombe Valley.

When I caught up with him 15 minutes later, he had joined a party of chimpanzees who were in the middle of a hunt. Reflecting on the sequence of events, I felt that the pant-hoots from above the falls had clearly conveyed to Frodo that a hunt was happening, or was about to start, which accounted for his rushing off. Pant-hoots are given in many contexts without significant acoustical differences in the call (Mitani and Brandt 1994).

Frodo's immediate reaction may have been analogous to your being in a forest with a friend who is walking in parallel 50 meters away. Suddenly your friend calls out "Snake!" In that first instant you may look down at your own feet, even though you quickly realize that the snake could hardly be anywhere near, given your friend's distance from you. Frodo's response may have been an instantaneous reaction of "Where are the colobus?" followed by the realization that the hunt was some distance away.

There is no evidence that Gombe chimpanzees, in order to solicit hunting aid from other chimpanzees, use a hunting call or other unique vocalization to indicate the presence of colobus or of a hunt in progress. Such a call is reported from at least one other study site, Kibale (Wrangham, personal communication). Of course, this sort of communication does not necessarily imply higher intelligence. Calls can be programmed to elicit responses without any higher cognitive skills at all, as in songbirds. But when calls are used in certain contexts, the voluntary, manipulative nature of such communication is strongly suggested.

On 3 July 1992 the chimpanzees had descended from their nests in trees in Kasakela Valley near a group of waking colobus. Though it seemed certain that the chimpanzees would hunt, dawn came and went and at 0800 the chimpanzee party headed up the slopes of Kasakela. I stayed with the colobus in order to obtain some census data. About 20 minutes later I noticed that one male chimpanzee, Beethoven, had remained and was sitting at the edge of a thicket staring up at the colobus, which were in a very tall *Ficus* tree over the streambed. Moments afterward, Beethoven performed a charging display against the buttressed roots of the *Ficus,* pant-hooting loudly and drumming his feet against the trunk. From up the slope came Frodo's distinctive wailing pant-hoot in response. Two minutes passed and Beethoven repeated the display and pant-hoot. By this time the colobus were excitedly alarm calling from their perches high above. The second pant-hoot brought another response from both Frodo and other males in the departing party. Frodo, according to the research assistants who had left with him, responded to Beethoven's second pant-hoot by calling back, then wheeling around and charging to the spot where Beethoven sat waiting. As Frodo and the others reached the tree they immediately climbed it and began hunting. Sixty-five minutes later the hunt ended, Frodo having caught a juvenile colobus, which he shared with both Prof and Beethoven.

This incident appears to have been a solicitation of hunting aid by Beethoven. Frustrated by the departure of the party with so many colobus close at hand, he managed to bring his companions back, whereupon they hunted. In the end he himself received a scrap of meat. Again, only the context or some unknown acoustical aspect of the pant-hoot could have indicated to Frodo the intent of the call, though in this instance the presence of the colobus would have been a clue to Beethoven's intentions.

ANOTHER EPISODE involving Frodo occurred on 17 September 1994. I was on Dung Hill early in the morning searching for W colobus group and listening for chimpanzee pant-hoots. At 0810 pant-hoots came from KK5, a ravine just south of the Kakombe waterfall that winds up and over the ridge to Mkenke Valley. After a short climb I reached KK5 and the ripening stand of mgege. Its grape-like pulpy fruits are sought by birds, monkeys, and chimpanzees. The chimpanzees were in one of the trees on the east side of the ravine hunting a group of colobus. Though five adult male chimpanzees were present, only Frodo was hunting; the others sat passively below watching the action. Frodo, having seen a mother colobus carrying her newborn infant, targeted her in spite of aggressive advances on him by three adult male colobus.

The hunt moved into a large bare *Albizzia,* and Frodo clung to one branch while the three colobus defenders clustered on a branch just above. The female colobus and her baby sat on a third branch slightly above and beyond the male colobus. For 20 minutes Frodo tried and failed to find a way past the males; the colobus had effectively screened him from his quarry. He then reached up, and after a few tentative shakes of the branch above him on which the colobus sat, began to whip the branch up and down. At the same time he bristled his hair and bounced up and down on his own branch. The entire central portion of the *Albizzia* shook as though in a windstorm. It was too much for the male colobus; the whipping forced them off the branch. As they leaped away, Frodo charged up and at the female, whose baby he snatched as the three male colobus regrouped and lunged at him. Frodo had the baby and nimbly swung down out of the *Albizzia,* leaving the colobus behind. All the while, Freud, Beethoven, and Tubi had sat watching from below; begging, they approached Frodo as soon as the kill was made.

Frodo is in many ways the most interesting of the Gombe male chimpanzees, having risen meteorically in the 1990s from an overgrown preadolescent to a large and powerful adult. Frodo's hunting talents emerged early in adolescence before I had the opportunity to watch him. By age 15, he was killing more colobus than any adult male; by 16, he was the catalyst for most hunts in which he was present, each year catching and killing up to 10 percent of the entire red colobus population within his hunting range. As Frodo matured he emerged as the most fearless predator in the community. The Tanzanian researchers whose experience went back to the hunters of the 1970s could remember no male more skilled.

Chimpanzees have individualistic hunting styles. Frodo is unintimidated by any colobus counterattack. Wilkie, while alpha male, was rarely in the thick of the hunt, but his status allowed him to take kills impunity from other hunters. In order to measure the effects of individuals on hunting success, I compared the influence of each hunter's presence on the success rates of hunting parties in which they were or were not members—all the while controlling for other confounding factors such as the overall size of the party and the total number of hunters present.

Frodo (Figure 4.10) was indeed the most influential of the Kasakela chimpanzees in his effect on the success rate of parties in which the different adult males were present. His brother Freud was second.

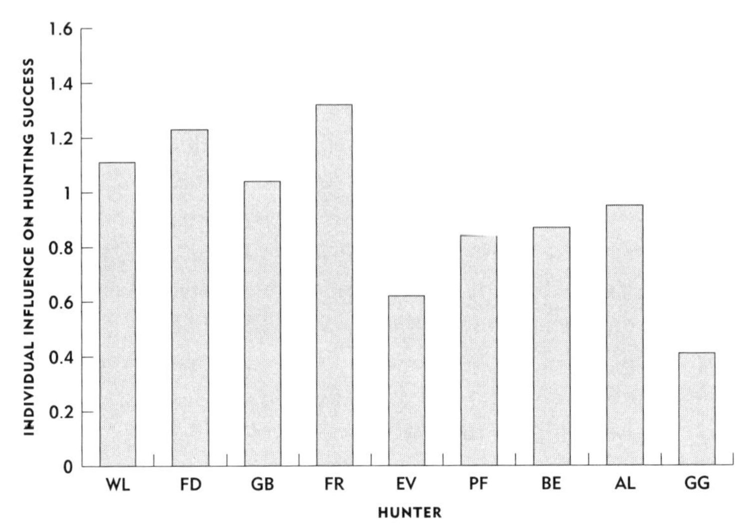

FIGURE 4.10. The influence of different males (plus female GG) on the rates of hunting party success at Gombe, calculated as the percentage of participation in parties divided by the success rate of the hunting party when each individual was present.

Frodo was credited with a kill 69 times between 1990 and 1995, including 50 times from 1990 to 1992. Other hunters, such as Evered, were less often in hunting parties and had less effect on the outcome of the hunt when they were present. The female Gigi is included in this analysis; as the only female who hunted often during the study, her impact approached the lowest male scores.

An additional measure of hunting prowess is how effective a hunter a chimpanzee is when hunting alone. Figure 4.11 shows that when no other male was present, the hunters differed dramatically in their likelihood of catching colobus. Both Frodo and Goblin were able to make kills at a high success rate without other male hunters in the party. The implication is that, even alone, either of these males is capable of hunting and killing a colobus monkey. This feat is impressive when the target of the hunt is being defended by a group of adult male colobus. Some of the other male chimpanzees, including Wilkie, made no kill when alone. These findings closely reflect my impressions of the males' relative abilities as hunters.

The breakdown of each hunter's tendency to capture colobus of different ages and sexes is also very revealing (Figure 4.12; Appendix 10). Whereas newborn infants (less than 3 months old) compose only 8

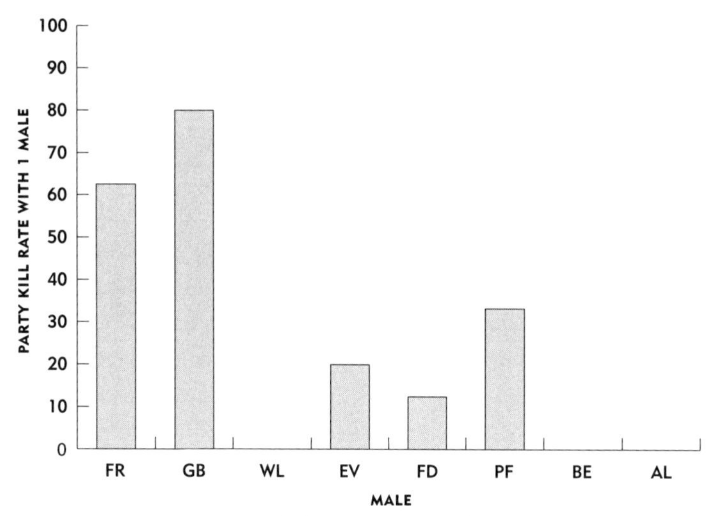

FIGURE 4.11. The influence of different males on hunting success rates of parties at Gombe containing only one male. Blank columns indicate zero success rates.

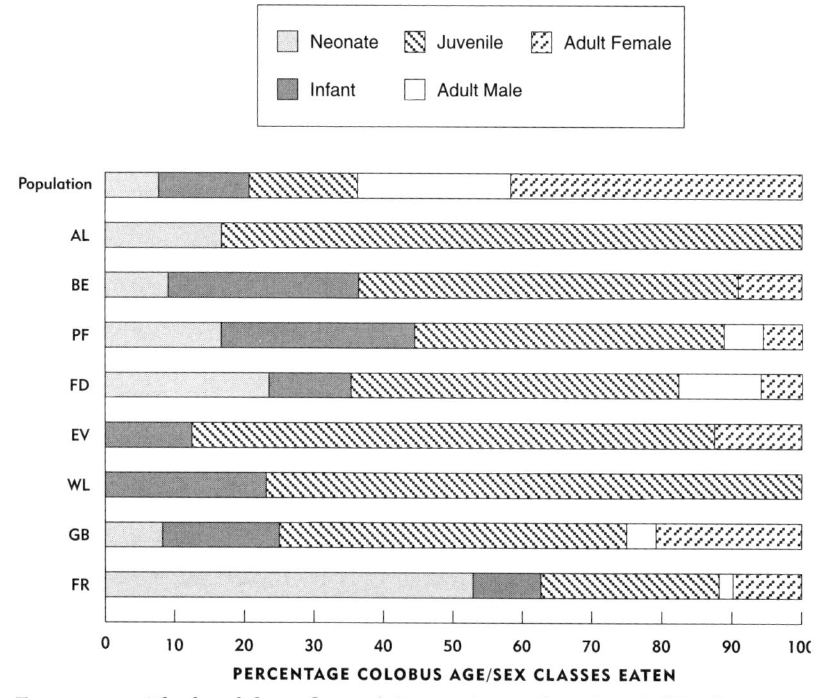

FIGURE 4.12. The breakdown for each hunter (more than three kills) of the age and sex of his red colobus prey. Compare these with the percentages of each age/sex class in the total Gombe red colobus population (top bar).

percent of the colobus study population, they represented more than one-half (52.9 percent) of Frodo's identifiable kills. Freud's colobus diet also included many neonates (23.5 percent), while the other hunters killed neonates roughly in proportion to their availability in the colobus population. The most startling statistic was the high percentage of kills that were juvenile colobus. Some hunters killed mostly juveniles (Evered 75 percent, Wilkie 76.9 percent, Atlas 83.3 percent), and all killed them in greater proportion than their number in the population.

The reason is very likely the difference in behavior when attacked between juveniles and infant red colobus. Neonates are carried, so they must be plucked from their mother's arms, which involves tackling not only the mother but also the males she stays close to during the attack. Older infants may be traveling independently, but they retreat quickly to their mothers in time of danger, so are also afforded maternal protection. Juveniles are independent enough to be alone during crises but not experienced enough to employ the appropriate behavioral responses to being hunted (and no doubt to other sources of danger as well). They tend to remain alone and unprotected on exposed branches at some distance from adult males, presumably the reason they are caught so often. The three hunters who relied mainly on juveniles as prey were the same three whose hunting agility and eagerness I considered to be poorest.

The effect of one especially fierce hunter such as Frodo may be felt throughout the local prey population for many years. If Frodo were to die or depart the Kasakela community while in his prime, the impact of hunting on the colobus population would be markedly reduced. By exerting such a strong influence on the colobus, Frodo may even affect the local forest ecology. If red colobus disperse seeds or crop new leaf growth through their combined feeding and ranging, then a decrease or increase in their numbers caused by hunting may have long-term consequences for both the prey and the habitat (Figure 4.13).

For decades the chimpanzees who lived to the north and to the south of the Kasakela community were a subject of much discussion and debate among chimpanzee researchers. How large were these neighboring communities? Had females from the Kasakela community, having gone through puberty and emigrated from the main study area, joined these unstudied communities?

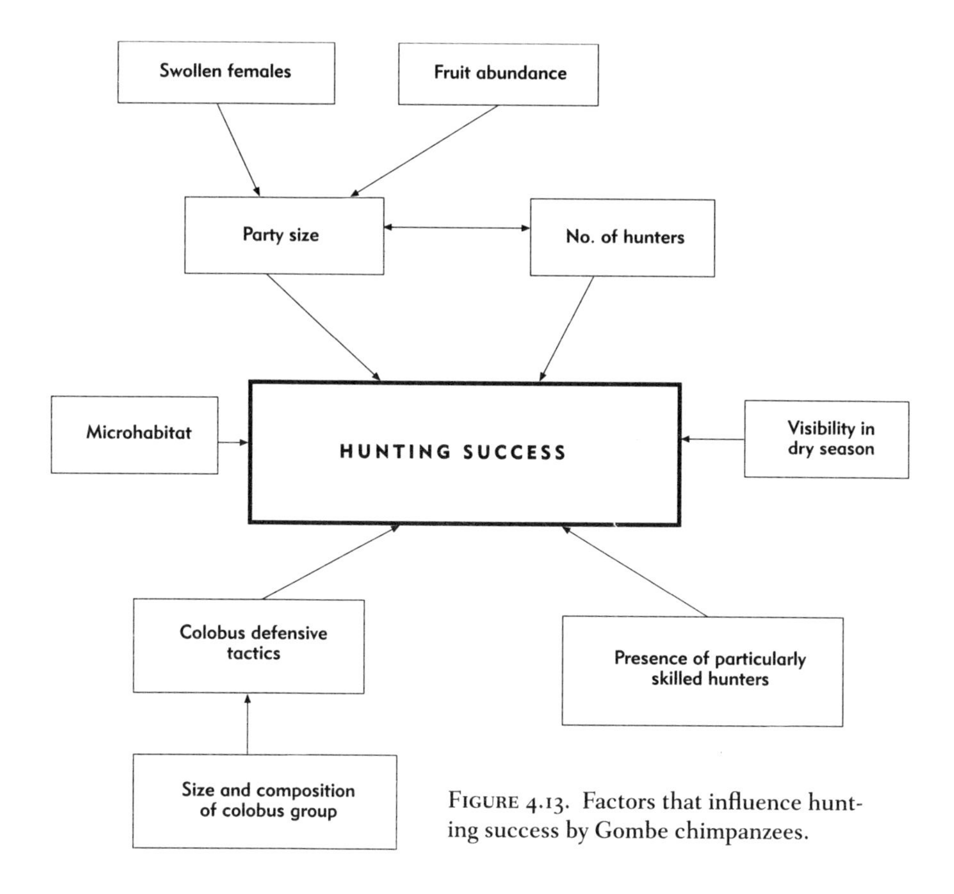

FIGURE 4.13. Factors that influence hunting success by Gombe chimpanzees.

Habituation and observation of the northern Mitumba community has been under way since the 1980s, and today we know a good deal about them. This is a small community of about 24 animals (reduced from more than 30 by an epidemic in 1996), including only 3 adult or adolescent males. The Mitumba chimpanzees occupy a small range of about 10 square kilometers sandwiched between the Kasakela community range and the northern edge of the park. Females who have transferred from the Kasakela to the Mitumba community, and vice versa, are now being studied as they pass through their life cycles in two different chimpanzee communities.

An interesting aspect of the Mitumba community is its lack of hunting. In spite of the hundreds of hours spent by research assistants in following these chimpanzees through Mitumba and surrounding valleys, no hunting was seen until 1993, and as of 1996 only a handful of hunts for red colobus have been witnessed. This apparent lack of interest in hunting by the Mitumbans was occurring a few kilometers away from the Kasakela chimpanzees, who were killing hundreds of red colobus and other animals every year.

This dramatic intercommunity difference in meat-eating may be explained by any of several causes, or a combination of thereof. First, we know from other field studies of chimpanzees that until the animals are quite habituated, hunting is rarely observed; it may be more frequent than it initially appears. At Ngogo in western Uganda, Ghiglieri (1984) considered the chimpanzees to eat meat very rarely, although more recent research has revealed that the nearby Kanyawara chimpanzee community may hunt as often as Kasakela chimpanzees do (Wrangham, personal communication). It appears that the lack of predation reported by Ghiglieri was simply a result of his not being able to spend long periods directly observing on the then-unhabituated Kibale chimpanzees. Since the Mitumba chimpanzees can now be followed for hours at a time and hunting is still not often seen, it appears that they do not hunt frequently (though the ratio of hunts to chimpanzee-colobus encounters is not known).

The same factors discussed earlier in this chapter that promote hunting opportunities and hunting success in the Kasakela community may be at work among the Mitumba chimpanzees too. Party size, number of males, and party composition may be as important in Mitumba as in Kasakela. The Mitumba community has only a small number of males, small party size overall, and perhaps lacks a particularly avid hunter such as Frodo to catalyze predatory behavior. In the Kasakela community, parties containing only two or three males hunt infrequently when they encounter red colobus (Stanford 1994a; Stanford et al. 1994b). Finally, the Mitumba colobus may be different, either in their behavior toward chimpanzees or in their foraging patterns, so that they are less often in contact with chimpanzees. Confirmation of these ideas will be pursued over the next decade as we learn more about differences, both cultural and ecological, between neighboring communities at Gombe and elsewhere.

Scavenging

The scavenging behavior of wild chimpanzees has been often discussed and sometimes misunderstood. In general, chimpanzees very rarely eat dead animals they have not killed; a large percentage of carcasses eaten by Gombe chimpanzee that they themselves did not kill have been pirated from other predators such as baboons (Morris and Goodall 1977). In 36 years of observation at Gombe, only a handful of records document true scavenging behavior—in which a party of chimpanzees has encountered a mammalian carcass and regarded it as a food item rather than an object of curiosity (Teleki 1973; Goodall 1986). This passive scavenging (in which a dead animal is found), is the sort that many researchers believe early humans performed during prehistory, along with the active pirating of kills from other predators. At other chimpanzee study sites, scavenging behavior has been recorded at a similar low frequency (Hasegawa et al. 1983).

Two arguments can be made for the importance of scavenging by Gombe chimpanzees. One is that its occurrence at all is noteworthy. The second is that scavenging at Gombe (and at Mahale) may be rare partly because of the lack of many species of ungulates, whose fawns provide savanna scavengers such as some baboon populations (Strum 1976) with most of the carcasses on which they feed (Moore 1996). When chimpanzee parties encounter a carcass in the course of their daily travels, they usually treat it with curious interest, perhaps including limited consumption of the dead animal. They rarely act as though they have discovered a highly prized food source.

I have witnessed this behavior only twice in the field, plus once in a lengthy videotape made by Eslom Mpongo. In November 1992 a chimpanzee party was crossing upper Mkenke valley when it stumbled across a partial carcass of a colobus, probably left over from one of the earlier kills. The remaining chunk appeared to be from an adult and was at least a day or two old. None of the adult males or female paid any attention to it, but the juvenile male Kris took it up a tall tree, where he sat gnawing on the dried meat for several minutes before discarding it.

The second scavenging episode is intriguing. On 13 January 1988 Eslom was following a mixed-sex party that included Frodo, Tubi, and Gimble, when they encountered the freshly killed carcass of an adult female bushbuck high in Kakombe Valley. According to Eslom, the carcass still had a good deal of recently dried blood on it and was intact except for the abdominal cavity, which had been eviscerated and emptied of tissue. All evidence pointed to a leopard kill, perhaps during the previous night. Here was an enormous package of meat, entrails, blood, bone marrow, and other body tissues, rich in protein and fat, available at no cost of foraging, risk of injury, or other obstacle to consumption. The response by the males present was to pause for about 20 minutes, examine and pick at the carcass, groom it a bit, and then depart. The females stayed after the males had left; Gigi, Wunda, and Kidevu showed greater interest than the other females but did not consume the meat. Gigi poked at it and groomed it, then turned the open carcass into a plaything, crawling into and rolling around inside the body cavity. She eventually left the scene without consuming more than a morsel herself. Throughout the incident, during which various chimpanzees approached the carcass, their reaction was one of curiosity and of recognition of a potential food item—but they did not feed (Muller et al. 1995). Scavenging may have a cultural element, in which encounters with carcasses are so infrequent that the chimpanzees do not recognize them as food. Mangoes, introduced to Gombe, are an example of a highly edible food that the chimpanzees ignore. But the long history of encounters with carcasses indicates that a culture of foraging for dead animals has had adequate time to develop; Gombe chimpanzees nevertheless do not regard dead meat, whether fresh or decaying, as desired food.

In this chapter I have reviewed only the basic patterns of chimpanzee hunting behavior and discussed some of the evidence for patterns that are not fully understood. On a given day, one or more of these factors will tilt toward the hunter or the hunted and make all the difference in the outcome. What remains to be examined is how these patterns affect the behavior and population biology of the colobus, and whether the interaction can explain why chimpanzees hunt.

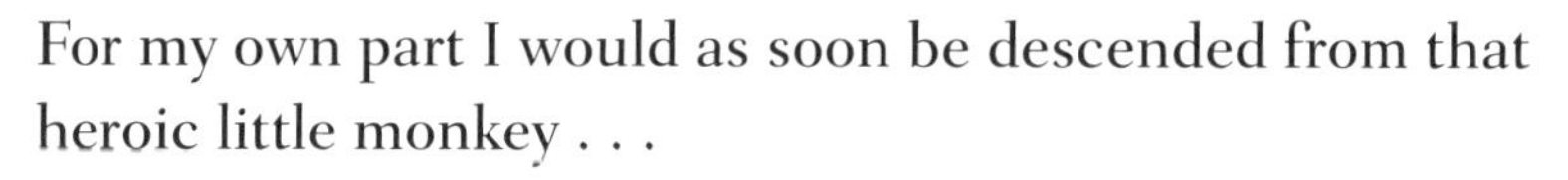
For my own part I would as soon be descended from that heroic little monkey . . .

—Charles Darwin, *The Descent of Man* (1871)

5. Red Colobus Monkeys as Prey

To most primatologists the colobine monkey subfamily to which red colobus belong is notable for two features: a highly specialized digestive anatomy and the propensity of males of some species to kill their rivals' infants. Primate field study has focused less on colobines (the colobus of Africa, the langurs and leaf monkeys of Asia) than on other Old World monkeys, even though they occur in roughly the same geographic areas and exhibit equal species diversity. This lack of attention is due to the social behavior of most colobine species, which does not feature the same frequency of struggles over sex or dominance status as that of their cercopithecine relatives (macaques, baboons, mangabeys, and guenons). Nevertheless, the colobines exhibit a range of ecological and social adaptations as diverse as those of any other mammalian subfamily.

The colobines are represented in Africa by three species complexes, the evolutionary relationships of which are unclear. Two of them, the red colobus and the black-and-white colobus, have been highly successful in colonizing a variety of African forest habitats and can occur at very high densities. The red colobus is found in forests from extreme western Africa through the equatorial portion of the continent and on the island of Zanzibar off the Tanzanian coast. Unlike other colobines, red colobus (and the related olive colobus) feature genital swellings in cycling females; for this reason some taxonomists place red colobus in their own genus, *Procolobus.* In Kibale National Park, Uganda, red colobus reach one of the highest densities and biomasses of any monkey species anywhere: nearly 300 animals per square kilometer (Struhsaker 1975). The black-and-white colobus complex is found across equatorial Africa and is usually divided into at least four separate species (*Colobus guereza, C. angolensis, C. satanas,* and *C. polykomos;* Oates 1994). The small olive colobus (*Procolobus verus*) is found only in coastal lowland forests of western Africa and is the most divergent of the African colobines. Estrous female olive colobus display a pronounced perineal swelling and sometimes carry neonates in their mouth (Oates 1994).

The red colobus occurs mainly in riverine forests in numerous disjunct populations. This distribution may have been caused by Pleistocene era climatic drying that left a once-contiguous species range fragmented (Struhsaker 1981; Kingdon 1989). The result of this long separation has been subspecific differentiation of red colobus into at least 13 separately classified forms (Kingdon 1971). The most divergent is the Zanzibar red colobus (*C. badius kirkii*), a beautiful animal whose red dorsal color is offset by a white abdomen and a white ring of plumed hair around a black face. Most of the other subspecies differ only in the pattern of the coat colors.

The red colobus population that occurs at Gombe is *Colobus badius tephrosceles.* It has no common name other than the local Kiha language name, *chondi,* and is not red at all. Gombe red colobus are dark gray above and buff-colored on the lower legs and undersides. The crown of the head is chestnut brown and the hands and face are black-skinned. the colobus is built like most arboreal monkeys: long-legged and lean-bodied, with a long tail that is used for balance during spectacular leaps between tree crowns. Red colobus travel via a combination of quadrupedal running and leaping along tree limbs.

Like the other colobine monkeys, red colobus possess only a rudimentary thumb. Adult males at Gombe weigh about 12 kilograms, based on my own estimate and comparison with known red colobus weights elsewhere, while adult females reach about 9 kilograms. When sexually receptive, adult females have a small perineal swelling that resembles a bright pink rosebud. Some males feature a remarkable form of female mimicry: subadult males, approximately the same size as adult females, often possess a pink swelling of the genital area that resembles a female swelling. This feature was first reported by Kuhn (1972) for red colobus of western Africa. In some other animal species, young males use their lack of adult male traits or even possess female-like traits. These traits may reduce aggression that young males would otherwise face from older and larger rivals. The *tephrosceles* subspecies of red colobus also occurs in both Mahale and Kibale national parks. Like red colobus over most of their range in Africa, *C. b. tephrosceles* is widely sympatric with chimpanzees in eastern Africa. Among the long-term chimpanzee study sites in that area, in only a few, such as Budongo Forest Reserve in western Uganda, are red colobus not also present.

Red colobus are one of the most common and observable large mammals in many African forests. They are also one of the best-known African forest monkeys, with nearly a dozen long-term studies having been conducted (Appendix 11). They live in social groups containing 4–14 males and 4–20 or more females plus immatures. The most comprehensive study of this species was conducted in Kibale National Park between 1970 and 1987 by a team led by Thomas Struhsaker. Most of the behavior patterns by which we characterize this species today were first described by Struhsaker (1975, 1981; Struhsaker and Leland 1985). He reported a lack of territoriality, females dispersing from groups while males remained, male dominance hierarchies, and infanticide.

The red colobus vocal repertoire was first studied by Marler (1970) and later by Struhsaker (1975) in the Kibale Forest. In most respects the use of vocalizations by Gombe red colobus resembles that of Struhsaker's Kibale population of the same subspecies. Table 5.1 compares vocalizations given by red colobus at Gombe and Kibale. At least four calls differ in context. One of these, the wheet, is a high-pitched, reedy whistle that is used by Gombe colobus when climbing into a tree crown away from approaching chimpanzees. At Kibale this context has not been reported, but Struhsaker observed its use to initiate group travel in the absence of chimpanzees. The call's function may therefore be quite similar at the two sites. I use the same names for the calls as Struhsaker (1975), after comparing the calls by watching Kibale colobus for several days in 1995 and also listening to calls on a tape provided by Struhsaker.

The first study of Gombe red colobus was done in 1969 and 1970 by Timothy Clutton-Brock of Cambridge University. He spent 10 months studying the feeding ecology, diet, and ranging behavior of red colobus occupying the same valleys in which some of my own study groups lived two decades later. Clutton-Brock (1973, 1974, 1975) compared the feeding and ranging patterns of red colobus at Gombe with those in Kibale, identifying food preferences and habitat use in relation to the

TABLE 5.1. *Selected vocalizations of Gombe red colobus, and differences in context from the same calls in Kibale National Park, Uganda (Kibale calls based on Struhsaker 1975 and personal observation). Both populations are assigned to the subspecies* C. b. tephrosceles.

VOCALIZATION	CONTEXT(S) AT GOMBE	CONTEXT(S) AT KIBALE
Wheet	Low-intensity alarm call, both sexes, often as animals move away from source of alarm but before giving chist calls	By males during intragroup aggression, before group movements, before grooming bouts, and before copulation
Chist	Contact calls by adults, mostly males; low-intensity antipredator alarm call to chimpanzees, humans, eagles, and any other source of alarm	Similar, but also intragroup tensions, by females to males, by males during displays, and during intergroup encounters.
Shrill squeal and prolonged squeal	Given by infants and juveniles when being attacked and captured by chimpanzees	Shrill squeal by immatures when threatened by adult aggression; prolonged squeal when infant is deprived of social contact
Quaver	During intragroup conflicts and especially when being hunted; high-intensity call when males attack chimpanzees	By adult males to copulating pairs, as harassment
Sneeze	Sneezing (very frequently given)	Sneezing and possibly other functions

availability and spatial distribution of food trees. While it was a short inquiry by today's standards, no previous field research had collected information on the feeding behavior of wild monkeys while simultaneously quantifying their resource base. This study set the stage for longer investigations of primate feeding ecology, especially those focusing on food choice in relation to plant defenses.

Male-Bonded Monkeys

On the morning of 28 July 1995 Frodo is foraging alone in the uppermost section of KK2 in Kakombe Valley. At noon he arrives at a towering

Albizzia tree in which a large group of colobus are feeding, and under which I have been sitting all morning. His wailing pant-hoots have been approaching for the past hour, but on reaching the colobus tree he stops and gazes intently overhead. Above, female and immature colobus feed watchfully while seven adult males give alarm calls—high-pitched chists in a frenzied chorus. After 10 minutes Frodo leaves the trail, pushes through the undergrowth to the buttressed roots of the tree, and begins to climb. All seven male colobus immediately descend the Albizzia to the lowest fork, where two massive horizontal limbs extend outward. They focus their collective attention on the ascending Frodo, alarm calling and leaping in place nearly shoulder to shoulder. Frodo is forced to stop and wait just below the fork, 2 meters below the cluster of colobus males. He scans the branches above and beyond them, perhaps looking for the infants that are his most frequent targets. After a 12-minute standoff, Frodo retreats to the ground and walks slowly away, glancing over his shoulder at the male colobus high in the tree.

The social system of the red colobus is unusual among colobines; it is multimale and also male bonded. It was formerly believed that nearly all primates, including colobines, live in groups of related females, from which males emigrate at sexual maturity. This conclusion was based on a small number of well-studied species. Today we know that a diversity of dispersal patterns exists and that in many species females migrate instead of (or in addition to) males (Moore 1984). In most red colobus populations that have been studied, females transfer between groups, leaving a nucleus of males who bear a presumed genetic relatedness to each other. Red colobus males in Kibale form dominance hierarchies, within which there are cooperative alliances that participate in intergroup encounters and in counterattacks against chimpanzees (Struhsaker 1975; Wrangham, personal communication). Because females transfer between groups and males generally do not, the core of the red colobus group is males who tend to be related to one another, although supporting DNA evidence is not yet available.

When confronted with the risk of predation, male colobus can choose group defense or crypticity to avoid detection. Male bonding in defense of the group may be promoted when cooperation is a kin-selected benefit, and when such cooperation is successful in reducing the risk of predation. Males tend to defend offspring whom they or their relatives are likely to have fathered, and are apt to defend females who have mated with them already, or may in the future. They usually support each other in dangerous circumstances, since each male in red colobus society is probably least distant kin of the other group males. When attacked by chimpanzees, red colobus males often launch aggressive counterattacks. Males may leap alone or jointly onto the backs of chimpanzees, biting them anywhere their teeth can find purchase, but also placing the colobus at grave risk from the canine teeth and grasp of the chimpanzees.

The Red Colobus Study Groups

I studied two colobus groups in particular: W group (for waterfall, the Kakombe waterfall being the approximate center of the group's home range) and J group (for *juu,* meaning "upper" in Kiswahili, because initially I thought—incorrectly—that Upper Kakombe Valley was the center of their home range). Tables 5.2 and 5.3 describe the members of the two study groups. Age categories are based on the following definitions:

Adult male: Recognized by his size, bulk, and visible genitals (though the scrotum is not apparent). Large canines. Body weight in excess of 10 kilograms.

Subadult male: About the same size and dimensions as an adult female and therefore difficult to distinguish in unhabituated animals. Has a pinkish patch on the rump, which at a distance resembles the clitoris of a female red colobus.

Adult female: Obvious nipples, clitoris visible at close range, and when cycling has a smallish pink perineal swelling similar to the first pubescent swelling in some cercopithecine monkeys. Otherwise similar to subadult male.

Subadult female: If nipples or clitoris are visible, can be distinguished at some distance from subadult male, but this is not often the case. Both weigh an estimated 8 kilograms. Age range from 2 to 4 years (females mature more rapidly than males, so spend less time in this age class).

Juvenile: Age range from 15 to 24 months, with wide variation in growth rates. Very difficult to sex reliably; unknown animals simply labeled immatures, though Struhsaker (1975) reported visible clitorises in older juveniles. Travel and forage independently of their mother and usually do not return to her immediately upon hearing alarm calls. Body weight range, based on one specimen, 2 to 4 kilograms.

Infant II: Older infant, about 4 to 15 months; more or less adult coloration. Travels independently of the mother, particularly after about 6 months old. Returns to her mainly to nurse, to sleep, and in time of danger. Body weight, based on one specimen, 1 to 2 kilograms.

Infant I: Neonate, less than 4 months old; coloration distinctly darker than that of older animals. Usually carried by the mother during group travel. By 2 months old plays and socializes away from her. Body weight, based on one specimen, 0.5 to 0.75 kilograms.

It is relatively easy to identify habituated monkeys, because at close range many idiosyncratic features become visible. Scars on the face, torn ears, broken toes, and broken tails are common, and adult males in particular have distinctive battle marks. If the animals are approachable, facial features can be learned and distinguished.

J group (Table 5.2) ranged across much of the lower valley that was also the Kasakela chimpanzees' core area. Their home range extended from the southern slopes of Kakombe below the ridge called Sleeping Buffalo, across the floor of the mid-lower valley, and up the northern slope to the region between two landmarks, the Peak and Bald Soko (Figure 5.1). Between these two high points were the fingers of several korongos up which J group traveled regularly in search of the leaves of the tall emergent *Albizzia* and *Newtonia* that grew on the korongo banks. When I began observing this group in 1991, it contained 24 animals: 5 adult males, 11 adult females, and immatures. This is roughly the mean size of colobus groups at Gombe, but only about half the size of colobus groups that live outside the chimpanzees' main hunting grounds.

Colobus monkeys do not engage in sex, fighting or dominance behavior on a minute-to-minute basis. Dominance patterns are more subtle and only become apparent after hundreds of observation hours. In J group an adult male I named Jason was the dominant animal, and he displayed this dominance during intragroup squabbles over food, during encounters with other red colobus groups, and when attacked

TABLE 5.2. *The members of red colobus group J in December 1991.*

J MALES	CHARACTER-ISTICS	J FEMALES	CHARACTER-ISTICS	IMMATURES	CHARACTER-ISTICS
JS (Jason)	High-ranking prime male	JN (Kink)	Dominant female, as large as smaller males		
JK (Jackson)	Old, heavy-bodied, high-ranking	JP (Tip)	High-ranking, missing last third of tail	JF, daughter of JP	Infant II
JU (Juma)	Young adult, high-ranking tail-tip missing	JG (Gray-back)	Pale dorsal hair	JH, daughter of JG	Infant II
JT (J-tail)	Low-ranking, split ear and broken toe	JR (Triangle)	Oddly shaped ears	JJ, son of JR	Infant I
JZ (Juzi)	Young adult, low-ranking, thin and dark	JE (Eagle)	Young adult or subadult	JD, son of JE	Juvenile
JB	Subadult	JA (Rabbit)	Broken toe, some-what peripheral	JY, daughter of JA	Juvenile
		JM	Mother of infant I	JO, daughter of JM	Juvenile
		JK	Distinctive face		
		JQ	Mother of infant II	JI, son of JQ	Infant II
		JC	Not reliably distin-guishable from JW		
		JW	Not reliably distin-guishable from JC		

by chimpanzees. He was clearly dominant over the other males and the females, based on his frequency of supplanting all other group members while feeding, but this dominance was only apparent during analysis of long-term data.

In J group in 1991 were four mothers with infants. Triangle (JF4) and JF9 each had small infants about 2 months old, while Tip (JF2) and

Table 5.3. *The members of red colobus group W in December 1991.*

W MALES	CHARACTER-ISTICS	W FEMALES	CHARACTER-ISTICS	IMMATURES	CHARACTER-ISTICS
WB (Buddha)	High-ranking, extremely heavy-bodied	WC (Clio)	High-ranking, prime	WE, daughter of WC	Infant I
WN (Bones)	High-ranking, prime, bony face	WP (Pineapple)	Pale buff legs, broken toes	WG, son of WP	Infant II
WK (Black)	Very dark dorsal hair	WL (Lucy)	Very red, long body, distinctive face	WT, son of WL	Infant II
WR (Bird)	Low-ranking, smaller	WM (Mango)	Small and low-ranking		
		WA (Apricot)	Distinctive face and tail kink	WW, son of WA?	Juvenile

Grayback (JF3) each had older infants, about 1 year old. Although it appears that only female red colobus migrate between groups at Gombe, I was never able to confirm that the females of the two study groups derived from different groups. Similarly, the adult males of the group were presumed to be kin inasmuch as no male was known to have transferred during a 5-year period. This could not, however, be known with certainty. J group was so frequently attacked by chimpanzees that tracking the development of offspring and the migration patterns of females was made difficult simply by the rapid turnover in group membership caused by predation. Nevertheless, in the course of five field seasons the size of J group changed little, and adult membership changed by only one animal.

Whereas most red colobus groups at Gombe are large and contain many males, females, and immatures, *W group* contained four adult males, five adult females, and four immatures (one juvenile and three infants). It ranged over a small area of the floor of Kakombe, across Dung Hill to the north, and on the southern slopes and korongos. Much of this range overlapped that of J group's much larger home range; a small part of W group's range also overlapped that of a third group, C.

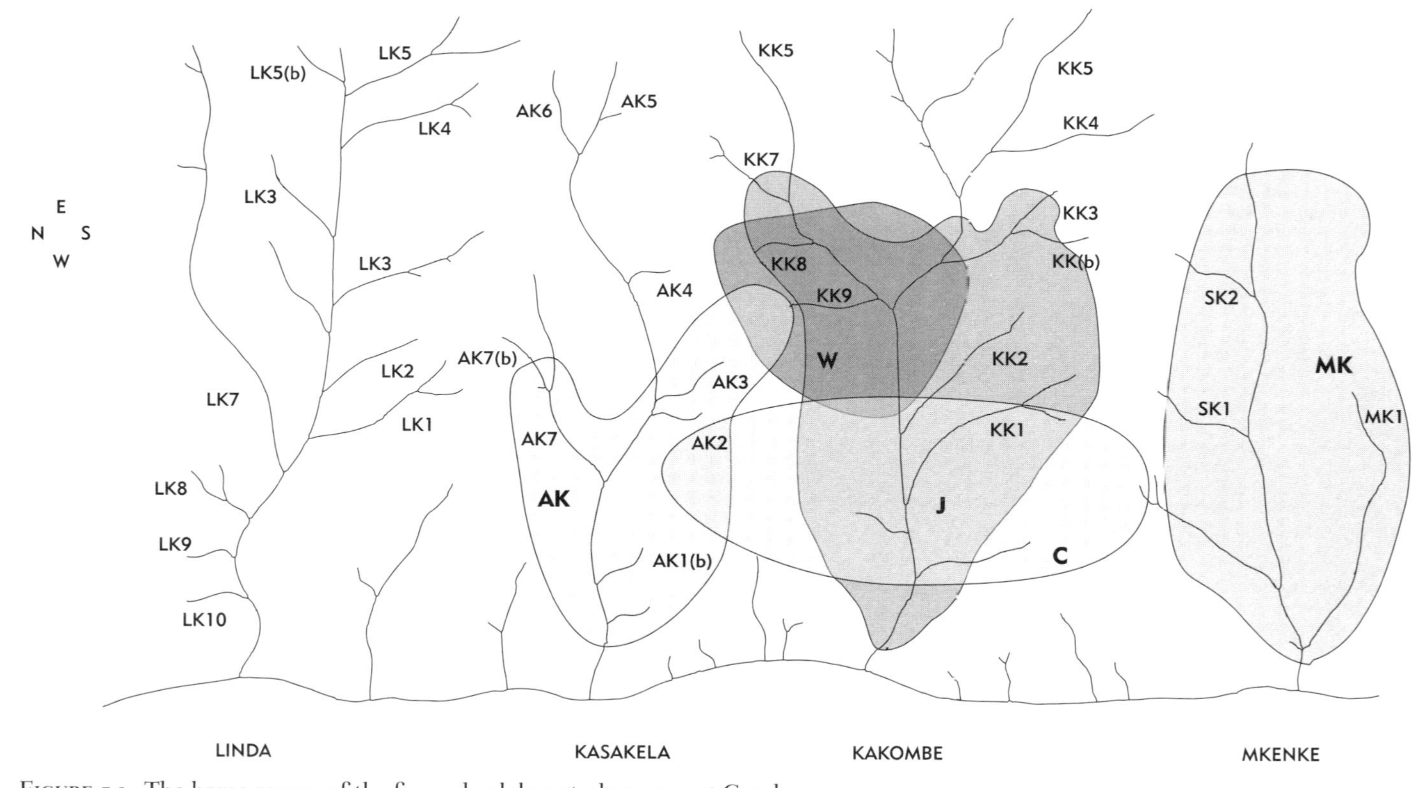

FIGURE 5.1. The home ranges of the five red colobus study groups at Gombe.

The highest-ranking male in W group I named Buddha, because he was the largest and heaviest red colobus I ever saw at Gombe. He probably outweighed each of the other three males of the group—Bones, Black, and Bird—by 30 percent or more. Buddha maintained a forceful dominance through belligerence toward the other males and the group's females from 1991 through 1993. In spite of their small numbers and small home range, W group put up fierce opposition during intergroup encounters with other local colobus groups, and this aggression was usually led by Buddha. The reason for the small size of W group is likely to have been continued predation by the chimpanzees. During the time I watched this group, it declined from thirteen animals to eight, after which it disappeared altogether.

At least two other red colobus groups were seen frequently in Upper Kakombe Valley. Above the waterfall, Kakombe Stream is rimmed by steep overgrown slopes, and tall emergent trees border the rushing stream. I spent little time in this area but sometimes encountered unknown colobus groups. I could not be sure that they were not the same groups I saw elsewhere. These groups ranged into the topmost forest patches of the park, within 50 meters of the rift escarpment itself. They therefore had a range that covered the upper 500 meters of elevation in Gombe. Groups appeared to cross over the ridge between Kakombe and Mkenke and enter the home ranges of the Kakombe study groups. In general, red colobus home ranges at Gombe are not demarcated by the valley ridges, even those that are topped with expanses of open grass.

In addition to J and W groups, I monitored three other colobus groups: C, MK, and AK (Appendix 12). They were identified only by the presence of two or three distinctive animals in each group. Scars, missing tails, torn ears, and in one case a spinal deformity, allowed me to identify them reliably. After I had censused these groups for several months, I was able to locate them by knowing their approximate home range and their location when last seen. *C group* was named for the frequency with which it appeared at the feeding station in the forest clearing, formerly called the camp. About three-quarters of the sightings of colobus groups within 50 meters of camp were C group. It had an elliptical home range that extended across the lower (western) parts of the ridges of Sleeping Buffalo and the Peak, covering parts of Mkenke, Kakombe, and Kasakela valleys. This configuration gave them greater access to the beach than the other colobus groups had. (Colobus groups in other parts of the park come to the rocky parts of the shoreline to forage and drink.) C group had 20 members in 1991.

MK (Mkenke) group ranged throughout the lower stretches of Mkenke Valley to the immediate south of Kakombe. It was the only group known to use the lower parts of Mkenke. However, in the upper parts of the northern slope of Mkenke lies a beautiful hanging valley named Chihaga, where colobus are hunted frequently. Some of these groups may have also ranged into lower Mkenke. These are the colobus visible from the public footpath that leaves the beach in Mkenke Valley and ascends the rift escarpment.

Finally, *AK (Kasakela) group* lived mainly in the Kasakela Valley, favoring the tall, emergent growth along the forks of the dry streambed in the valley bottom. AK group also ranged up and over the Peak and the Upper Peak and onto the upper northern slope of Kakombe, thus overlapping to some extent with J group.

Daily Life

Monkeys are most often associated with treetops, and the red colobus is no exception. As dawn breaks, red colobus groups are usually sitting in small clusters of three or four in the crown of a tall (more than 30 meters) tree, very often *Albizzia* or *Newtonia*. They may have spent the previous evening in these trees, and now breakfast begins there. Most feeding is at the tips of branches. Some group members will climb to the highest part of the tree crown to capture the sun's first warming rays as they break over the rift. The colobus often spend midmorning traveling through the canopy along a streambed, or up and down a valley slope, feeding as they go. From midday until late afternoon the monkeys forage and sleep. Another active feeding bout late in the day lasts until dusk. By day's end the group has typically traveled between 400 and 800 meters (seldom more and often much less) and has spent most of its time 15 to 30 meters above the ground. The colobus will have fed on 15 to 20 plant food species, met no other colobus group and experienced no major upheaval within their own group (Appendix 13). They have seen or heard, on average, one potential predator.

I SPENT ONLY 4 or 5 days a month following red colobus groups from dawn until dusk. More often I conducted half-day follows in order to devote time to chimpanzee parties as well. On 87 days on which dawn-to-dusk travel was recorded, J and W groups moved a combined average of 393 meters (range, 0–850 meters). In an average half-day the animals moved only a short distance, and when feeding in clumps of preferred foods they sometimes did not move at all. Colobines are known to banquet on abundant leafy foods rather than forage selectively for higher-quality food. When necessary, however, colobus can interrupt foraging to travel swiftly, leaping between tree crowns and down vertical slopes, and covering great distances within a few minutes. This happens mainly when they are alarmed by a potential predator or an approaching human.

Group Size

The size of a colobus group is the outcome of the decisions of individual members to live in groups that are smaller or larger, because of benefits received as a result. Such benefits can be predation related; it may be better to live in a big group to defend against predators or to spot them approaching. Alternatively, it may be better to be in a small group to avoid detection in the first place. Optimal group size may be food related, since many mouths mean more competition for limited resources. (Possibly, larger groups are better at defending food patches once they have been located).

The mean size of red colobus groups at Gombe is 28 (± 6). This figure, though comparable to most other red colobus populations, masks considerable variation. Group size varied from 13 to 69, and all groups larger than 45 were recorded only from the far northern and southern portions of the study area. Mean group size within the core hunting area of the Kasakela chimpanzees was 18.7. Kibale red colobus studied by Struhsaker had a mean group size of 34 (CW group) and similar variation, from 9 to 68.

Only two known populations have varied substantially from Gombe and Kibale figures. The group size of red colobus at Tana River, Kenya, averaged 18 individuals and varied from 12 to 30 (Marsh 1979). The large size of Gombe red colobus groups reported by Clutton-Brock in the late 1960s is intriguing: his main study group contained 82 animals, a number larger than any colobus group recorded today in his former study area. The mean size of the five groups he monitored was 55 (Clutton-Brock 1973, 1975), about twice the size of the average group inhabiting the same area 25 years later. These data could be taken as evidence that the ecology of Gombe has changed, or even that chimpanzee hunting is a recent phenomenon and that the group size of Gombe red colobus was once much larger than it is now. Another possible explanation is that group size has been affected by cyclic patterns of predation. During the 1960s, when chimpanzees at Gombe were provisioned and spent long hours at the feeding station in Kakombe Valley, they may not have been hunting colobus as often as they do today.

A social animal can often minimize its risk of capture by staying close to other group members. Such an action may simply reduce the numerical odds that this animal, rather than a neighbor, will be the object of the predator's attack. The hypothesis, referred to as the Geometry of the Selfish Herd (Hamilton 1971) may account for the grouping patterns observed in a wide variety of animals in nature. Red colobus, however, must always make a Hobson's choice. A prey animal can either be very cautious about being exposed to predation and therefor forego access to high-quality food patches, or take risks in order to have better access to such food.

Gombe red colobus live in cohesive multimale groups that persist over long periods. Yet the hour-to-hour foraging structure of the group is different. Colobus at Gombe have a group spread that is very wide. The mean diametrical distance from one edge of J group to the other is 49.3 meters (N = 86 samples of group spread, 1991–1993; s.d. ± 10.5), and many times the edge-to-edge distance was so great that I could not locate all individuals or subgroups. This characteristic did not show significant seasonal variation. Groups traveled and fed in a loose spatial association. The behavior is very different from that of other colobine monkeys, such as black-and-white colobus or many langurs, who tend to remain in highly cohesive groups throughout the day.

The biggest groups had the strongest tendency to be dispersed, perhaps because of the greater mathematical likelihood of wide dispersion in large groups. Groups sometimes spent the entire day apart but would be seen together the following day. Without individual identification it would be easy to mistake two groups foraging in close proximity for one large group, and vice versa. Red colobus, in spite of other risks, apparently prefer to space themselves widely in order to maximize their individual access to food. And their important food sources may themselves be dispersed widely—a parameter that has previously been seen as the reason for noncohesive grouping in red colobus monkeys (Struhsaker and Oates 1975).

Habitat Use

Red colobus have been called riverine forest species (Davies and Oates 1994) because they tend to be found along river courses. At Gombe they certainly fit this pattern, in that they spend a large percentage of time in the emergent trees that line the banks of the park's streams. Their preference is for the two large Mimosaceae trees of that habitat, *Newtonia buchanani* and *Albizzia* sp., whose leaves together composed well over half of the diet in some seasons (Appendix 14). Other habitat types fragmented are woodland of the lower valley slopes, miombo on the higher slopes, and occasionally grassland (Figure 5.2). Use of grassland is hard to understand because colobus travel on the ground with an awkward amble that is not well designed to escape terrestrial predators. If surprised on the ground more than a few meters from a tree, a colobus could not possibly escape. Nevertheless, on a number of occasions I saw J group on the ground, including on the Peak, when chimpanzees were calling not far away. Clearly, the differential use of these habitats by colobus may affect their vulnerability to chimpanzee hunting, since the chimpanzees themselves use the same habitats seasonally according to where their important foods are located.

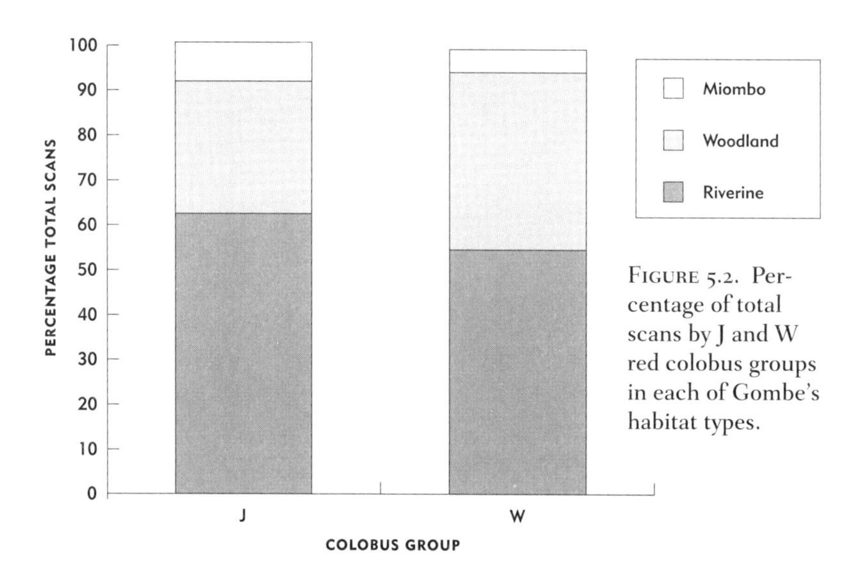

FIGURE 5.2. Percentage of total scans by J and W red colobus groups in each of Gombe's habitat types.

The sizes of the home ranges of the five red colobus groups varied between 40 hectares and 1 square kilometer, though all fell within the range of variation reported previously for red colobus. A positive correlation between the size of the colobus group and the size of its home range (Figure 5.3; $r^2 = .46$) was consistent with the pattern found for most group-living primate species. Home range should be related to group size, since larger groups need more space in which to locate resources (Clutton-Brock and Harvey 1977). In addition, the home ranges of other local groups may impact one another, leading to constrictions and expansions of home ranges. In nonterritorial species whose home ranges overlap, multiple groups share resources and use parts of their ranges differentially. The home ranges of J, W, and C groups overlapped extensively in Kakombe Valley. Even so, by August 1994 a new group (NK) appeared in the same areas used by J and W; these 45 animals were using the former range of W group through August 1995.

Diet and Feeding

Gombe red colobus eat a variety of plant foods, although their diet is monotonous compared to chimpanzees in the same forest. Whereas chimpanzees have been recorded eating at least 184 plant food species (Wrangham 1975), of which about half (47 percent) are fruit, red

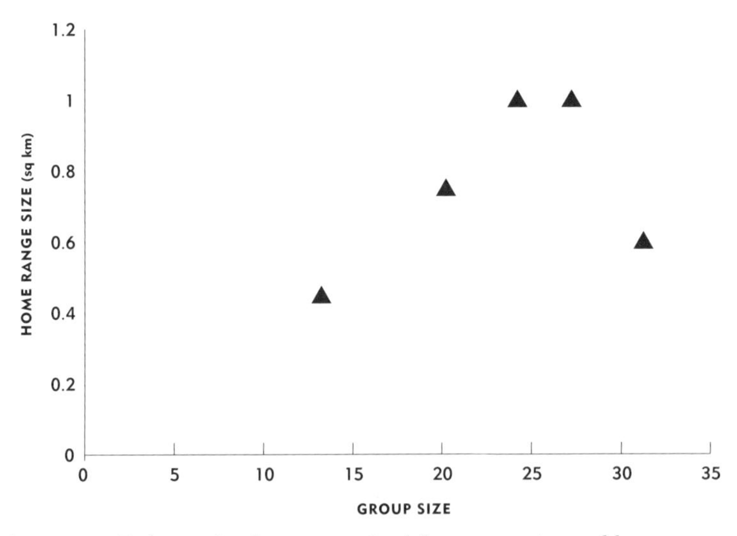

FIGURE 5.3. Relationship between red colobus group size and home range for the five study groups at Gombe.

colobus use only about 65 species (the major ones are shown in Appendix 14). This lower diversity has much to do with colobus concentration on just a few abundant sources of leaves that are available for most of the year. Chimpanzees, on the other hand, travel far and wide in search of widely dispersed ripe fruits.

As with other leaf-eating primates, colobus choose young leaves over mature ones; being tender, they are more easily digested and contain a higher proportion of nutrients. Figure 5.4 shows that new leaves are preferred over mature leaves during most of the year. Fruit is a prime source of carbohydrates for a wild monkey, and ripe fruit especially is high in sugars. Colobine monkeys typically avoid ripe fruit, whose acidity may disrupt their digestive system (Kay and Davies 1994). But fruit is eaten in greater quantities than one might suspect. While ripe fruit is a rare dietary item, unripe fruit is eaten heavily from July through December. When colobus were seen feeding in a tree that contained ripe fruit, they seemed to be selecting the unripe fruits in the same crown. For example, *mtobogoro (Ficus vallis-choudae)* is a large fig that grows mainly along the banks of streams at Gombe and is a preferred food of both colobus and chimpanzees. The two species eat different parts of the plant, however. When both species feed in the same mtobogoro at the same time, colobus are usually eating

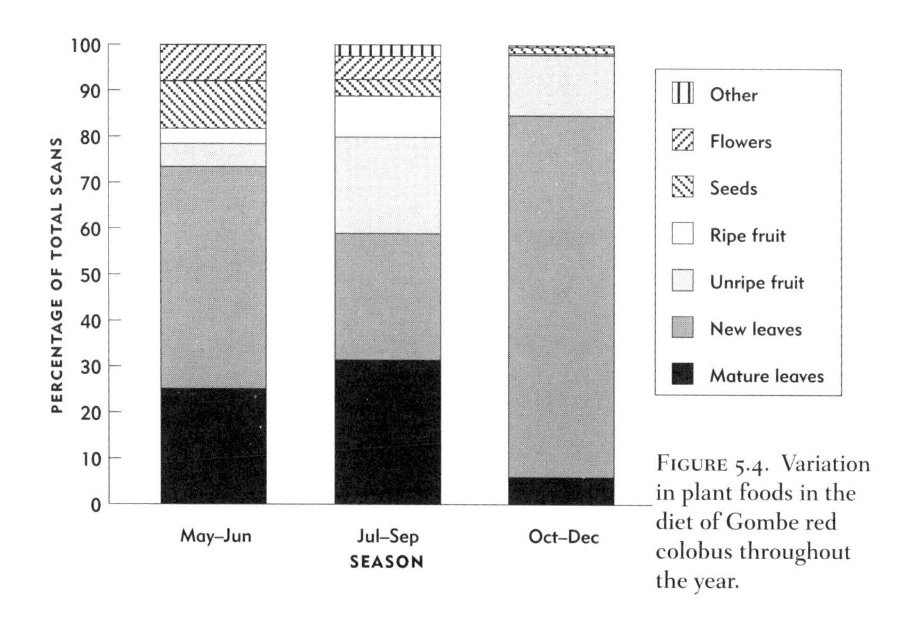

FIGURE 5.4. Variation in plant foods in the diet of Gombe red colobus throughout the year.

unripened fruit while chimpanzees (after a brief sniff that presumably indicates ripeness) are selecting the ripe fruit.

The colobus diet consists of a small number of widely available foods, but nevertheless varies seasonally. Appendixes 15 and 16 show the monthly variation in eating the five principal foods in the annual diet of the colobus. The two primary leaf sources, *msebei* (*Albizzia* sp.) and *mka* (*Newtonia buchanani*), were eaten in every month of the study (May through December). *Albizzia* consumption peaked in the middle of the dry season at the same time that *Newtonia* was eaten much less. The next three most important foods were mainly sources of leaves (*Sapium ellipticum*), fruit and seeds (*mrama; Combretum molle*), and flowers (*mninga; Pterocarpus angolensis*). Of these, *P. angolensis* and *C. molle* both showed strong seasonal patterns. Their consumption peaked in the late dry season of September and October, whereas the leaves of *S. ellipticum* are eaten mainly during the wet season.

If chimpanzees seek widely scattered ripe fruit and colobus forage for more evenly distributed leaf patches, the diet of the two species might seem to be entirely different. But in fact they share five of the ten plant food species most used by each. In addition to *Ficus vallis-choudae,* both red colobus and chimpanzees make important use of *mninga* (*Pterocarpus angolensis*) flowers during the month of September. *Mguiza* (*Pseudospondias microcarpa*) is a major source of fruit for chimpanzees in August through October; colobus eat mguiza leaves throughout the year. *Msulula* (*Pycnanthus angolensis*) fruits are eaten by chimpanzees; colobus eat msulula leaves. The pulpy grape-like ripe fruits of *Syzigium guineense* are relished by chimpanzees during the fruiting months of September and October. Colobus eat leaves and the new, unripe fruit of this species at the same time. When both species utilize the same part of the tree during the same season, they may encounter each other often as a result, leading to additional predation.

When two members of a monkey group meet in the tree in which they are feeding, each has one goal: to obtain the maximum amount of food. This goal may be thwarted if an animal feeding a meter away is able to take one's food away with impunity. So we should expect to see wild primates trying to prevent others from sharing whatever ideal patch of food they have located. This intragroup feeding competition may play a role in the lifetime reproductive output of a female monkey through repeated displacements from valuable food sources, leading eventually to malnourishment.

Fruits occur as nutrient patches that are often scattered widely in the forest, while leaves are less dispersed. Leaves should therefore be less often the objects of competition among individuals. Moreover, the level of competition ought to be higher in larger groups than in smaller ones because the number of possible competitors is greater. This was the case among the five Gombe red colobus groups (Figure 5.5). There was a marked difference between the largest (AK, 31 members) and smallest (W, 13 members) groups in the frequency of intragroup aggressive behavior that occurred in a feeding context. Feeding competition was defined as the number of aggressive behaviors per hour between two adults within 10 minutes before or after a feeding bout. Likewise, levels of feeding aggression were essentially the same whether the group was feeding on fruit or leaves. This fact could result from overall size; both groups were small compared with those outside the core hunting area of the chimpanzees. Alternatively, intense intragroup competition may not take place among red colobus at Gombe. They may thus fit the frequent colobine pattern of weak competition for food. Or these may be the wrong two groups on which to base any conclusion, because predation may maintain their size below the level at which intragroup feeding competition becomes important.

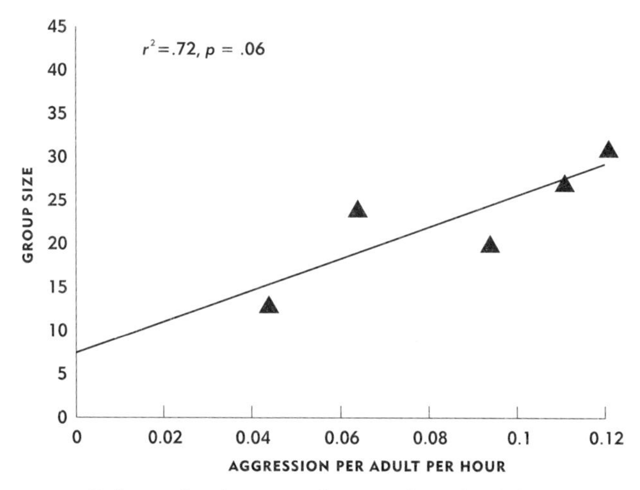

FIGURE 5.5. Relationship between the size of a red colobus group and the frequency of food-related aggressive interactions among adults of the five study groups at Gombe, 1991–1995.

Intergroup Behavior

To a monkey watcher, intergroup encounters are both exciting and informative. They are rare events, and the nature of the encounter may help to explain patterns of socioecology, such as territorial defense or the lack of it. They may also indicate the social dynamics within each group by revealing who participates, who does not, who takes an aggressive role, and who is a bystander. In some species males engage in intergroup encounters, while in other species only females do. Such confrontations also reveal whether or not territorial space is being defended, and whether one group is dominant over the other and perhaps why. They may even indicate the resource over which the encounters occur (for example, defense of food or defense of females) and thereby suggest the driving forces of the social system of the species. Lethal territoriality was, for example, a key piece of evidence showing that the social system of chimpanzees is based on male kin groups, whose degree of kinship corresponds well with their tendency to cooperate in defending territories and attacking intruders.

I saw only eight encounters between Gombe red colobus groups, all in 1991 and 1992 (Table 5.4). All of them involved at least one of the study groups, and four involved J and W groups together. The infrequency of intergroup encounters is puzzling because the home ranges of the study groups overlapped, in some cases extensively, and the density of animals was quite high. Mutual avoidance is likely, but is very difficult to distinguish from random failure to encounter one another in the forest. When groups did meet, there was sometimes a mutual disregard, but on most occasions the outcome was one group supplanting the other and usually by chasing between the males of both groups.

On 22 November 1991 the males of J and of W groups were in close proximity in the canopy along the banks of Kakombe Stream. Buddha, the dominant male in W group, repeatedly chased each of the five adult males of J group away from the stream area. They returned over and over again to chase the other W males (but not Buddha) back across the stream toward the south and to the base of Meat Ridge. Over a period of 90 minutes J supplanted W, which moved away to feed elsewhere. Most encounters were similar: much chaotic chasing without a definitive victor, followed by one group's retreat. Females played no apparent role but

TABLE 5.4. *Observed intergroup encounters at Gombe involving red colobus groups J and W in 1991 and 1992.*

GROUPS	DATE	LOCATION	FOOD PATCH	OUTCOME
J-unidentified	2 October 1991	KK dell	?	Unidentified group supplants J group
J-W	2 November 1991	KK3/Kakombe Stream	?	J supplants W after repeated chases by males of each group
J-W	18 June 1992	KK3/Kakombe Stream	*Albizzia* spp.	J supplants W
W-unidentified	8 July 1992	Lower KK6	*Albizzia* spp.	W supplants unidentified group from *Albizzia* as it passes through
J-unidentified	1 September 1992	KK2	?	Fight between group males; infant from unidentified group later seen dead
J-unidentified	4 September 1992	KK5	*Syzigium guineense*	J supplants unidentified group
J-W	23 September 1992	Lower KK7	?	W males chase J males and attempt to gain access to J females; J males chase off W males
J-W	31 October 1992	KK6	*Albizzia* spp.	Mutual avoidance

continued feeding, watching, and protecting their infants. This sex difference in participation may have been related to the immigrant status of all the known group females: they had less to gain from participation than did males with long-term and potentially kinship bonds.

The nature of the encounters, during which no fighting for access to females occurred among males, suggested that the confrontations were about access to food. During one encounter involving J group, an infant from the other (unidentified) group died. The circumstances were ambiguous, but it appeared that the infant was killed by a male from its own group rather than by an extragroup male. Infanticide, reported widely for at least one colobine species (the Hanuman langur, Hrdy 1977; Newton 1988), has been seen rarely in red colobus (Struhsaker and Leland 1985).

6. Before the Attack

On 3 July 1992, AK group awoke at 0630 in trees bordering Kasakela Stream. A few minutes later a party of 18 chimpanzees, including 9 adult and adolescent males, began to stir in their night nests only 25 meters and two tree crowns away. As dawn came, the colobus ate, groomed, and watched the chimpanzees. Although more than 30 minutes passed, during which the colobus could have moved away from the area, they did not attempt to do so. The chimpanzees eventually left their nests and departed without hunting, until Beethoven called them back. One juvenile member of AK group was captured in spite of a fierce defense by the male colobus.

Why didn't the colobus move away earlier that morning when they had the opportunity to do so? Are there aspects of red colobus behavior or ecology that mitigate the risk of predation? Why do male colobus join together to launch risky counterattacks on their hunters? The fearless counterattacking behavior of male colobus monkeys is among the most dramatic examples of apparent cooperation among primates. But how cooperative is it? The behavior of red colobus when attacked presents a fascinating opportunity to test hypotheses about the costs and benefits of male-bonding in nonhuman primates.

Studying Predation

When the risk of being eaten by a predator is a daily threat to a monkey, we would expect to see a response to this risk in the behavior of individuals and of groups. Antipredator response involves multiple behavior strategies, each of which can mitigate the risk of predation in a different context. But each strategy carries a cost that must be balanced against the benefits of additional safety from predators.

One low-cost way in which an individual or a group can respond to predation risk is by minimizing the odds of encountering predators in the course of daily ranging. The prey might avoid certain places where predators occur or hunt frequently; the cost will be feeding opportunities lost at that site, and possibly an additional travel cost. If predators are encountered, other strategies may be used, such as alarm calls to warn relatives and group mates. If the predators are not avoided or repelled before an attack occurs, the prey may be required to flee or to counterattack. Depending on the species and the context, both tactics can be high risk.

The type of antipredator tactic employed should thus depend on (1) the risk of attack—an open habitat is more dangerous where most predators are open-country hunters—and (2) the context-dependent risk, in which the benefits of counterattacking vary with the body size of the prey and the potential danger to the counterattacker. Predators, whether cats, birds of prey, or snakes, have their own patterns of habitat use, and we might expect that a primate group would (to the extent possible) avoid those areas most frequently used by the predators. An animal and its group should engage in high-risk behaviors, such as approaching a water hole, at times when or in places where the risk is minimal relative to the quality and quantity of food there.

There is little evidence that the risk of predation really keeps most primates away from any part of their home range (van Schaik and Mitrasetia 1990). Key resources such as ripe fruit trees and water holes attract not only primates but also many other species, all of which are fair game for predators. Among vervet monkeys studied in Amboseli National Park, the monkeys were eaten by large pythons in the marshy areas around water holes more often than in any other part of their home range (Cheney et al. 1988).

Intragroup social dominance may determine who is able to choose where within the group to be. Among the vervets, individuals at the front of the group are more likely to become victims, even as they are reaping nutritional benefits by finding food first. In brown capuchins, social status predicts which individuals are peripheral and which are central; the peripheral animals end up with less food and greater risk of becoming prey (Janson 1988). A long-standing debate exists on whether being in the center of a baboon group is better for avoiding the risk of being eaten (DeVore and Washburn 1963; Collins 1984).

To lessen its exposure to an attacker, an animal can adopt certain straightforward tactics that do not require drastic expenditures of energy or time. First, since most predators can attack only one animal at a time, the larger the group, the lower the odds that a specific individual will be attacked. Moreover, the geometry of the group as it feeds, rests, and travels may play a role in protecting each member; staying in a particular place may make that individual less vulnerable to attack (Hamilton 1971). At the edges of the group there are fewer eyes to spot an incoming eagle or the pounce of a big cat. Selfishly, therefore, a monkey should remain at the center of the group when the risk of predation is greatest at the periphery. There is evidence that this occurs in primates: Tamar Ron and her colleagues (1996) found that, for chacma baboons, being centrally positioned in the group was a benefit of being high-ranking. Failure to be in the center led to a greater risk of predation by leopards, a fact that has interesting implications for the social dynamics of the group. The result is members jockeying for position relative to one another, seeking access to food while using each other's spatial position to remain safe from the threat of predation.

Group Size and Movement

Natural selection tends to favor behavioral tactics that minimize the risk of a predatory encounter. If an encounter occurs, natural selection mitigates the risk of attack to reduce time and energy lost to each colobus who otherwise must continually fend off attacks from predators. A long-standing question in primate socioecology is whether it is better to live in a large or a small group, and why. In a large group each member is numerically less likely to be the victim of an attack, and with more eyes and ears to be on guard, the risk of an attack may also be lessened. And in species such as the red colobus, which aggressively counterattack their predators, more group members should be advantageous—especially more males who can serve as defenders of the other group members.

Carel van Schaik and Maria van Noordwijk of Duke University compared the behavior of long-tailed macaques living in a predator-rich forest in Sumatra with another population living on an island that lacked predators. They found that monkeys on the offshore island lived in significantly smaller groups, suggesting that the advantage of group size lay in detecting predators. In their absence, the optimal group size was smaller (van Schaik and van Noordwijk 1985). Other factors may have accounted for this difference, however. The geographic separation between the two monkey populations meant that they occupied different habitats, ate different foods, and perhaps even had somewhat different climates.

Large groups may be detrimental to individual safety if they are more easily detected in the forest. I found that I could much more easily locate a large group than a small one. Large groups tend to be noisier when traveling, because of the cumulative noise of many animals. When they are at rest, the odds are greater that at least one animal will be moving about and will reveal the group's location. Also, as colobus groups travel or feed, they spread out across the forest canopy. The spread is likely to be wider for larger groups, so that the odds of walking along a trail and spotting one portion of the group overhead is greater when the diameter of the group is also greater. These impressions are reinforced by the reports of other observers (National Science Council 1981).

Furthermore, intragroup food competition may place constraints on upper size limit (Terborgh and Janson 1986), while the ability of larger groups to control food sources may promote a minimum size limit (Wrangham 1980). Travel costs combined with feeding competition may also constrain the number of individuals who can forage together (Janson and Goldsmith 1995).

While there is a theoretical optimal group size for each set of local ecological factors, the optimum is affected by two paradoxical factors. First, groups cannot be optimal in size unless the individuals in those groups choose to stay or leave to enhance their individual fitness. If a group has a theoretical optimal size of 20, then immigrants are likely to be drawn to this group for its very attractive qualities—safe haven from predators and acceptable levels of food competition. Once several immigrants have arrived, though, the group size is no longer optimal, to the detriment of both residents and immigrants. Optimality should therefore be considered to exist over a shifting range of group sizes such that increases and decreases within some interval are acceptable to all the animals involved.

A second paradox is that however great the evolutionary benefits of living in a certain size group, proximate effects may mask any manifest stability. When predation is intense, the annual predation mortality may keep the population below the theoretical optimum and also below the food resource limitation level for the habitat. The result is that recorded prey group sizes reflect mainly the proximate effects of cropping by predators.

Among Gombe red colobus, the effect of group size on the likelihood of being attacked was clear. Larger groups were attacked significantly more often per encounter with chimpanzee parties (Figure 6.1). When a chimpanzee foraging party met a group of 45 colobus, they were nearly three times as likely to hunt them as they would a group of only 20 animals. This effect is linked to two factors: the number of adult males in the group and the number of immatures. Because Gombe chimpanzees show a strong preference for immature colobus monkeys, I predicted that groups with more immatures would be more likely to be attacked. This prediction was borne out (Figure 6.2). There was a nearly linear relationship between the number of infants and juveniles in a group that was encountered by chimpanzees and the likelihood of an attack. The average number of immatures per adult had no bearing on the relationship. Chimpanzees preferred to hunt when the number of immature prey was large.

Large colobus groups also tend to have more adult males than do small groups; an abundance of defenders might discourage hunting by

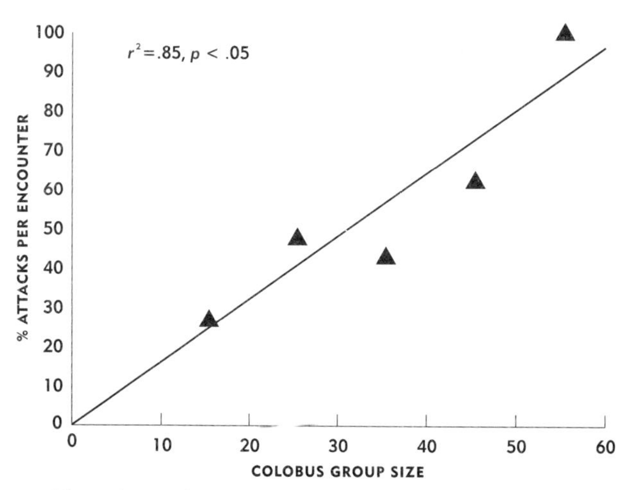

FIGURE 6.1. The relationship between the size of a red colobus group and the likelihood that it will be attacked by chimpanzees (Gombe, all encounters, 1991–1995).

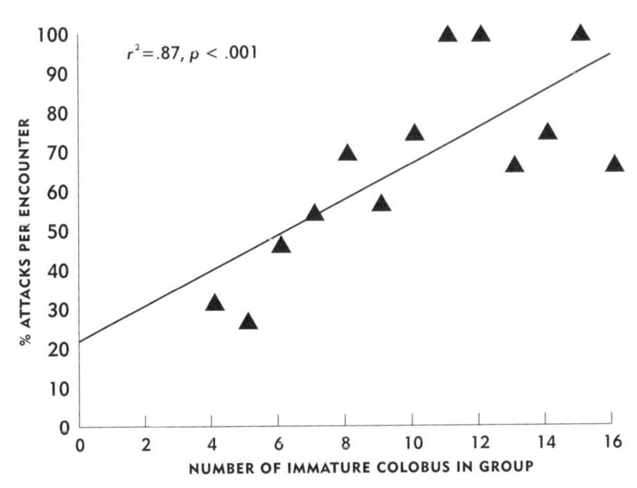

FIGURE 6.2. The relationship between the number of immature red colobus in a group encountered by chimpanzees and the likelihood it will be attacked (Gombe, all encounters, 1991–1995).

chimpanzees. Instead, there is a positive, but statistically insignificant, correlation between the number of adult males in the group and the rate of hunts per encounter (Figure 6.3).

Is the greater vulnerability to attack of large groups offset by advantages to individual animals or to specific age or sex classes? In any group of 50 red colobus, an average of 19 animals are immatures. Each immature therefore has about a 0.05 chance of being the individual attacked. In a group of 20, only 4 group members will be immatures, so the odds are 0.20. The group of 50 will, however, be attacked nearly three times more often, so the risk to an immature in that group is 0.15, or 0.05 lower than to an immature in a smaller group. This slightly lower risk of attack is bolstered by the more numerous male defenders in large groups.

Resident females in large groups may be at a marked reproductive disadvantage relative to other females who migrate to smaller groups with a lower risk of attack. We must therefore also consider what benefits females gain in large groups that prevent them from simply transferring out to avoid continual attacks from chimpanzees.

Although I focused more on the effects of predation than on feeding ecology, some information is available to weigh the costs and benefits of large group size. Large groups may be good or bad for the individual depending on local ecology and the abundance, nutritional content,

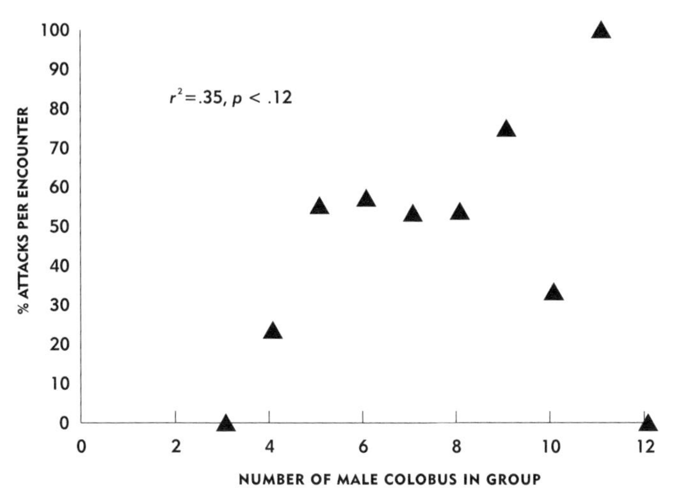

FIGURE 6.3. The relationship between the number of adult male red colobus in a group encountered by chimpanzees and the likelihood that it will be attacked (Gombe, all encounters, 1991–1995).

and distribution of food in the group's habitat. The disadvantage of large group size from a nutritional perspective is that the more competitors one has for food, the less food one may get. Other aspects of group living may interfere with spending as much time as desired engaged in feeding (Janson 1985). In bigger groups, food squabbles may break out more often, reducing the time spent eating. This possibility assumes that food competition within groups is more important than competition between groups. If the latter is the case, it might be beneficial to a female's reproductive success to live in a large group that can control food patches. A test of these competing theories of grouping patterns is whether (a) intragroup food competition is really stronger in larger groups, and (b) whether females in large groups have higher individual reproductive success than females in smaller groups.

Red colobus in large groups have more food-related aggressive interactions with their groupmates than do those in smaller groups (see Figure 5.5). In addition, colobus in larger groups spend less time feeding (Figure 6.4). These two pieces of evidence indicate a negative side to living in a larger red colobus group. Van Schaik (1983) examined measures of female reproductive success to see if females living in larger groups, and therefore presumed to obtain less to eat, actually suffered decreased reproduction. He estimated their reproductive outputs by

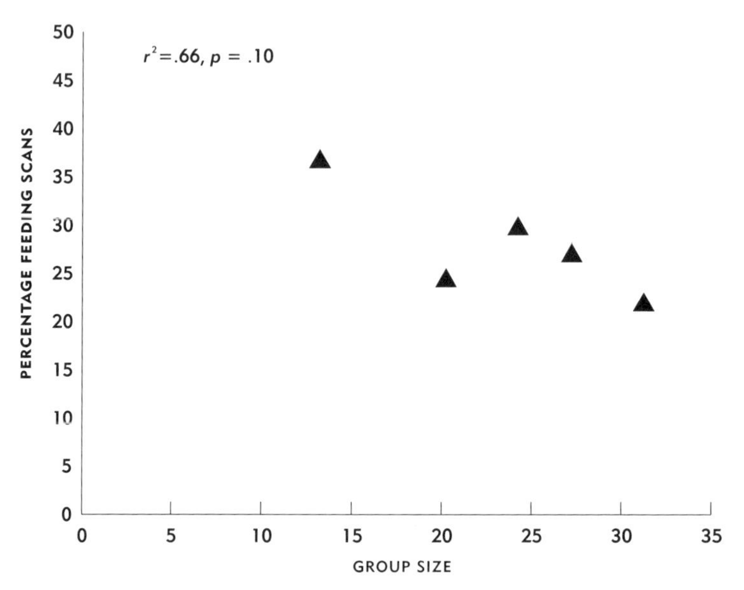

FIGURE 6.4. The relationship between the size of a red colobus group and the percentage of observation scans in which at least half the group is feeding, based on the five study groups at Gombe, 1991–1995.

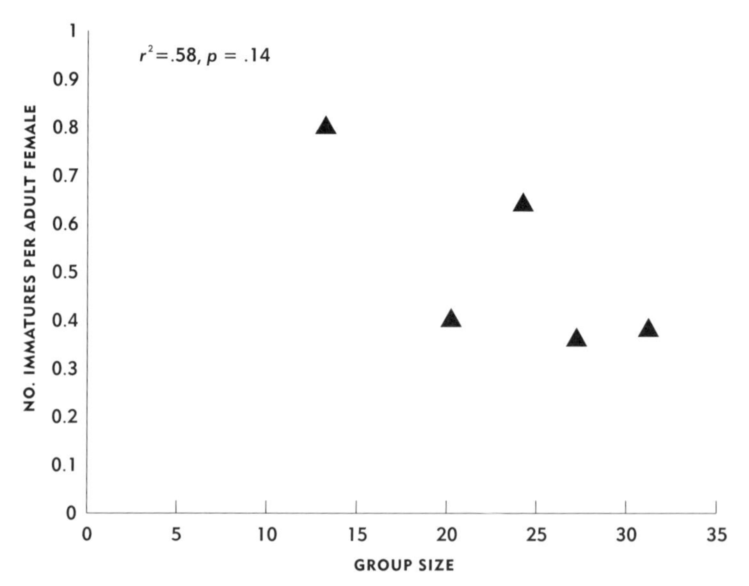

FIGURE 6.5. The relationship between the size of a red colobus group and the number of immature colobus per adult female in the group, based on the five study groups at Gombe, 1991–1995.

measuring the ratio of immatures to adult females who were the likely mothers in the group. Van Schaik found that females in larger groups had lower reproductive success.

Figure 6.5 shows that in the study groups there was a negative relationship between the size of a group and the number of immature colobus per adult female in the group (though this trend failed to reach statistical significance). Apparently these females suffer decreased fitness from food competitors. However, because female red colobus emigrate from their natal groups and sometimes leave their older offspring behind, this result must be viewed cautiously. It is unlikely that as more females join a group, their individual fitness increases, which suggests that large groups are not nutritionally beneficial to females. Following Wrangham's (1980) predictions about the behavior of groups toward one another, I expected that larger groups might displace smaller groups from feeding sites, and that this ability would be a key advantage for big groups. Intergroup encounters were rare enough that I never observed this behavior, however.

Red colobus groups suffer differently from predation depending on their size. Although living in a big group generates a higher risk for some age classes, countervailing influences may be present. If, for instance, a red colobus group of 40 animals were attacked once a week, whereas another group of only 20 were attacked only every two weeks, then it might be safer to live in the smaller group—assuming that the risk from migration out of a group is lower than the detriment of remaining in it. If the odds of being captured were reduced by half in the larger group, this circumstance might mitigate against wholesale migrations to smaller groups.

The distance and direction that a group travels in a day are the outcome of the need to find food, the degree of competition for food within the group, and the need to avoid being eaten. Charles Janson and Michele Goldsmith (1995) showed that length of daily travel is partially a function of food competition, in that animals who consume less food per distance traveled must travel farther each day in order to make up that deficit. This thesis assumes that group members live on a tightly constrained narrow energy budget. However, because groups can be attacked by predators, there may be ways for the group members to lessen the odds of attack—or at least increase the odds that they themselves will not be the likeliest victims. Red colobus may avoid being eaten by avoiding areas of frequent chimpanzee activity, or by positioning themselves in their groups during travel (or feeding) to minimize the risk of being grabbed by a predator.

Beyond the statistical dilution of risk achieved by living in large groups, spatial positioning and associations can be of critical importance. The Selfish Herd model does not, however, apply well to red colobus, in that the odds of attack and of dying are greater in large groups. Among Gombe red colobus, spatial position in the group changes quite markedly when chimpanzees approach (Figure 6.6). Males cluster together to prepare to counterattack the predators climbing from below. Females gather up their infants and attempt to stay near one or more adult males; they tend to reduce the distance from their nearest neighbor in the group (not including their own infants), who is most often a male. Juveniles do not respond to a threat by aggregating; they remain farther away from other group members than other age or sex classes. Thus they are caught very frequently relative to their representation in the population.

Vigilance

As social mammals feed they periodically scan their immediate surroundings. This vigilance behavior is found in a wide range of animals. Visual scanning of the environment is strong circumstantial evidence of the important effect of predation on the behavior of wild primates. Vigilance comes at a measurable price, which is the time spent scanning the environment for potential attackers rather than eating or engaging in other behaviors. Therefore, it benefits each animal in a social group if only some of the members are vigilant at any one time (Pulliam 1973; Caraco 1979). As the size of the group increases, then, each animal needs to spend less time scanning and can spend more time doing something else. So in large groups we expect to see less scanning behavior per individual. By implication the predation risk for each individual is also reduced, and mathematical models have supported this prediction (Pulliam 1973).

Vigilance is the most-studied behavior related to predation risk, though much of the work on vigilance has been done in nonprimates. Brian Bertram (1980) showed that ostriches tend to scan their surroundings less often as the size of their foraging group increases. Kimberly Sullivan (1985) studied downy woodpeckers to assess the benefits

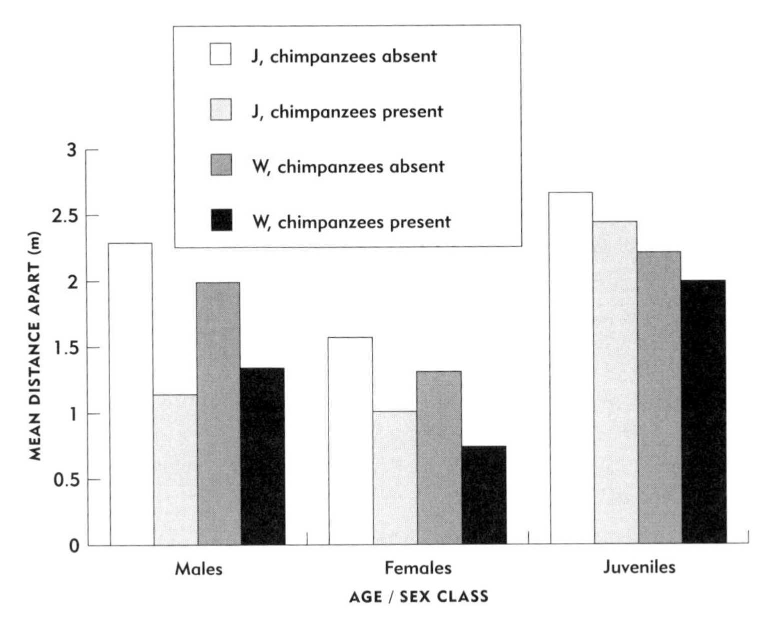

FIGURE 6.6. The mean distance to the nearest neighbor among the males, females, and immatures of J and W red colobus groups in the presence and absence of chimpanzees at Gombe, 1991–1995. Adapted from Stanford 1995.

of foraging in flocks and found that birds that foraged alone spent significantly more time scanning and less time feeding than those in flocks of three birds or more. Primate studies have also reported vigilance behavior. Jan de Ruiter (1986) found that Venezuelan wedge-capped cebus monkeys in small groups spent more time scanning than those in large groups. Treves (in press) reported the same finding for red colobus monkeys in Kibale National Park, and Mariel Baldellou and Peter Henzi of the University of Natal in South Africa (1992) made the same observation for vervet monkeys. The cost in feeding efficiency is presumed to be the critical factor preventing animals from being too vigilant. An animal who sacrifices a great deal of feeding will suffer a decline in nutritional intake.

Small-bodied primates may attempt to compensate for their inherently high vulnerability to predators by being especially vigilant (Heymann 1990). Carlos Peres (1993) showed that stable mixed-species groups of two tamarins, *Saguinus fuscicollis* and *Saguinus mystax,* were attacked nearly once a week by birds of prey. They coped with this threat by scanning intensively for predators. An interesting distinction was that *S. fuscicollis* scanned mainly for terrestrial threats from below, while *S. mystax* detected attacks by aerial predators from above. Since the predominant threat to both species was from the air, added vigilance may have accounted for the formation of these polyspecific groups through active instigation of *S. fuscicollis.*

A problem with nearly all studies of vigilance is failure to control for the many confounding factors. An animal that scans its environment cannot be assumed to be looking for predators; it might equally well be keeping an eye out for strange males (Rose and Fedigan 1995) or group-mates (Cords 1990). At the same time, engaging in any other behavior is done at the cost of time spent scanning (Cords 1995). Mark Elgar (1989), in a review of the vigilance literature for birds and mammals, concluded that nearly all of these studies fail to take into account alternative explanations for vigilance.

In few studies has the researcher tried to control for other confounding variables. Those most widely ignored are food distribution and density. If a group of monkeys moves into a food tree and the food density is high, the animals are likely to spend more time feeding than if the food is less dense and the animals have to move to another feeding location. Likewise, if the quality of the food is very high, the animals may spend little time scanning, because that time is feeding time lost. Both of these possible explanations for scanning and the lack of it have been demonstrated in bird studies (Barnard 1981, cited in Elgar 1989).

Competition for food within a primate group may also confound evidence on vigilance for predators. In feeding on the same food, group members are likely to increase their time spent feeding in order to maximize energy intake. This increase would be greatest in large groups and would result in lower rates of visual scanning. It would thus be unrelated to predation risk, while giving the appearance of being an effect of lowered risk in large groups.

Not only potential predators but also rival males from other groups and social partners within one's own group are prime targets of attention. This is the conclusion drawn by Lisa Rose and Linda Fedigan (1995) in their study of *Cebus capucinus* in Santa Rosa National Park in Costa Rica. They found that males were highly vigilant, but they interpreted that vigilance to be aimed mainly at detecting rival male cebus monkeys.

I collected information on scanning while watching J and W groups during the first two years of my study, then decided it was not a reliable measure of antipredator detection. Red colobus in the larger study group J scanned their surroundings less frequently than the adults of W group. I sampled 10-minute intervals throughout the day and found that the rate of scanning was about 50 percent greater in J group (2.55 times per minute for J versus 1.33 times per minutes for W; two-tailed *t*-test, $p < .01$). The significance of this finding is hard to interpret. In large groups the first priority of each colobus may be to obtain food; more time spent looking for predators means less time to spend eating. The rate of scanning in both groups was lower when the animals were feeding on leaves rather than fruit, and declined after colobus had been feeding for longer than 10 minutes on any type of plant food.

This finding suggests that the type of food eaten influenced scanning rates, as suggested by Elgar (1989). Perhaps when a colobus group arrives at a tree that is full of tender leaf flush, each animal's priority is to get a stomachful of leaves first, an urge that gives way to more cautious feeding behavior only after some minutes have passed. Why be less vigilant when eating leaf flush than when eating fruit? If new leaves are a preferred food that colobus do not want to sacrifice to their groupmates for the sake of safety, then it pays to eat more and scan less. Alternatively, if new leaves take time to ingest because the tender young growth must be selected from tougher, older foliage, the amount of time devoted to other activities may drop. The mean distance between group members was also slightly higher in J group than in W group, a fact that may have influenced the rate of scanning. Many uncontrollable factors may bias observations of inferred scanning for predators. Scanning is a useful detection strategy when predation is a continual risk, as it is for colobus, but the evidence that colobus do in fact scan to detect predators is equivocal.

In most studies of primate vigilance, males have been the more vigilant sex (de Ruiter 1986; Rose and Fedigan 1995; Cowlishaw in press). The number of males in a primate group varies from one to dozens, depending on the species. The number may be most strongly influenced by intragroup relationships, such as the ability of a male to monopolize the females. It may also be influenced by intergroup explanations, such as the competitive ability of larger groups to dominate and defend food sources.

Protection against predators is another compelling reason for multiple males. Van Schaik and van Noordwijk (1989) conducted an experimental field study using models of snake, eagle, and jaguar predators to test the reactions of Peruvian cebus monkeys. The researchers found that male cebus monkeys were highly vigilant for predators and also tried to drive predators (or at least models of them) away from the group. They reasoned that female cebus monkeys tolerate the presence of several males in a group because of the predator detection and defense that males provide for females and their young. Among African vervet monkeys, both Cheney and Seyfarth (1981) and Baldellou and Henzi (1992) found males to be more actively engaged in vigilance than are females. Baldellou and Henzi found that additional males and subordinate males are able to join or remain in the group not because they are desired by females—indeed, they become extra mouths competing for food—but because females are unable to exclude them.

At least one comparative study has attributed the number of males in a primate group directly to the risk of predation. Carel van Schaik and Mark Hörstermann (1994) compared monkey grouping patterns in Africa, Latin America, and Southeast Asia. They found that more males per group were present in species in Africa and the neotropics than in Asia, independent of overall group size. This finding corresponds to the presence of large monkey-eating eagle species in African and neotropical forests, and the lack of comparable raptors in South and Southeast Asia. The investigators suggest that predominantly one-male groups of Asian langurs and multimale neotropical howler groups may be constituted differently, owing to varying degrees of predation risk from large birds of prey. While a variety of ecological and phylogenetic factors appear to play a role in influencing the number of males in a group, risk of predation appears to have the principal effect.

Mixed-Species Groups

A group of red colobus are feeding in a large fig tree in Mitumba Valley when 15 red-tailed monkeys approach. The colobus ignore the approaching red-tails and continue eating leaves. The guenons walk through the center of the colobus group; one strolls up to a colobus female and presents its rump to be groomed. The colobus obliges. Later, while still feeding within a few meters of one another, the red-tail is approached by a young baboon, who slaps at the red-tail to chase it out of the fig tree. The female colobus rushes over and chases the baboon away, allowing the red-tail to feed in peace.

In many African forests guenons (genus *Cercopithecus*) travel in associations of multiple species. In East Africa, red-tailed monkeys (*Cercopithecus ascanius*) and blue monkeys (*C. mitis*) particularly commonly form a dual-species group. In the New World tropics, polyspecific groups are often composed of tamarins and marmosets (*Saguinus* and *Callithrix*), plus cebus (*Cebus*) and squirrel monkeys (*Saimiri*). Such groups are most often made up of the smaller-bodied primates found in each major geographic area of Africa and the neotropics. It is unclear why these groups form, since more mouths mean more intragroup food competition. They are not chance associations that arise simply because many animals are drawn to the same fruit tree, although establishing this fact in each case is necessary (Waser 1982).

In some tropical forests primates form mixed-species groups in which two or more species forage together for hours or days at a time. These foraging associations may confer both feeding and antipredator benefits (Terborgh 1990). The antipredation benefit is likely to be enhanced detection because of the extra ears and eyes the mixed group possesses (see for example, Boinski 1989). Some species in the group or flock may capitalize on the alarm calls of the most vigilant species. Peter Rodman (1973) and John Terborgh (1990) have pointed out that whereas African and Latin American forests have one or more large monkey-eating eagles (the harpy eagle in the New World and the crowned eagle in Africa), the forests of both Southeast Asia and Madagascar lack such a species. These forests also lack mixed-species primate groups.

A team of French researchers lead by Annie Gautier-Hion and Jean-Pierre Gautier have showed that for West African rain forest monkeys, mixed-species groups of guenons take on many of the characteristics of mixed-species flocks of birds. Different species in the supergroup assume roles that aid in predator deterrence and detection (Gautier-Hion et al. 1983). These investigators identified *Cercopithecus pogonias* as a sentinel species; by issuing a loud call, this guenon warned other monkeys in the megagroup of the approach of aerial predators. Foraging efficiency may also have been enhanced for all three species in association, but the researchers concluded that antipredator benefits were the primary *raison d'être*. The benefits of foraging together may also lie in greater ease of finding food (Gautier-Hion et al. 1983) or in the opportunity to capitalize on information held by other species about food (Munn 1985).

For Gombe red colobus, opportunities to associate with other species are limited to two arboreal monkeys that are common at Gombe: the red-tailed monkey and the blue monkey. These two guenons occur together in mixed groups and hybridize freely. Most groups contain red-tails, blue monkeys, and several hybrids of mixed parentage. Red colobus are seen frequently with both species; more than one-third of all observations of J group were made with either red-tails or blues in the immediate vicinity (Table 6.1). The mean duration of each contact was 36 minutes ($N = 448$ contacts). Red colobus groom red-tails and vice versa; Sharon Watt, during her doctoral study of red colobus socioecology, saw a red-tail masturbating a male red colobus! This relationship therefore involves social interactions as well as potential ecological advantages for each species. Identifying these ecological factors is not, however, an easy task.

To see whether having a larger group size with more eyes (and mouths) had an effect on either colobus vigilance or feeding, I compared the behavior of colobus when they were alone with the time during which they associated with red-tails, blues, and mixed groups of red-tails and blues. Red-tailed monkeys at Gombe occur at a density similar to that of red colobus (see Appendix 2), while blue monkeys occur at a lower density. When red-tails were in the vicinity of colobus,

TABLE 6.1. *Associations among red colobus group J and other arboreal monkeys at Gombe, 1991–1993. Association criterion = red colobus within 10 meters of other species.*

HABITAT	% SCANS WITH RED-TAILED	% SCANS WITH BLUE
Total	30.3 ± 3.6	7.2 ± 2.5
Riverine forest	34.3 ± 4.1	8.0 ± 2.8
Woodland	16.0 ± 2.9	1.2 ± 0.5
Miombo	1.1 ± 0.3	0.0
Grassland	0.0	0.0

the colobus were more vigilant than when alone (Table 6.2), suggesting that there was no predator-detection benefit afforded them by the red-tails. Further they may have had to spend more time watching for chimpanzees because the red-tails made their group larger and more detectable.

My earlier skepticism about vigilance data should be reiterated here. The presence of one or two other species may cause the colobus to scan the guenons themselves, accounting for what appears to be extra antipredator vigilance when the guenons are nearby. The same applies to colobus–blue monkey associations, though the sample was too small to show any strong effect and many colobus-blue associations took place when red-tails were also present. Red colobus also spent slightly less time feeding (though not significantly so) when guenons were with them. Rather than releasing the colobus from the feeding-time constraints imposed by the need for predator vigilance, association with red-tailed monkeys may not have provided the colobus with any opportunities to eat more food.

Both red colobus and red-tailed monkeys give alarm calls when they detect chimpanzees. My findings indicated that, more than half the time, colobus alarm calls preceded calls from red-tails when chimpanzees approached (Table 6.3). In a smaller number of instances in which colobus and blue monkeys were together, red colobus always called first. The colobus, however, rarely moved away from the source of the danger, either when alone or when with other monkey species. Instead, the two guenons fled quickly and were gone by the time a

TABLE 6.2. *Comparison of vigilance and its effect on feeding by red colobus at Gombe in mixed-species groups and alone. Paired* t-*tests, * = not significant, *** $p < .01$.

SPECIES	NUMBER OF COLOBUS LOOK-UPS PER 10 MINUTES	% SCANS COLOBUS FEEDING
Red colobus alone	2.55	29.0
With red-tails	3.96**	28.3*
With blues	3.13*	31.1*
With red-tails and blues	3.20**	28.7*

TABLE 6.3. *Detections of chimpanzees by mixed red colobus/red-tailed monkey/blue monkey groups at Gombe, 1991–1993. Includes responses to all sightings and calls of chimpanzees whether or not an actual encounter ensued.*

CHIMPANZEES DETECTED BY —	N	FIRST TO ALARM CALL	FIRST TO MOVE AWAY
Red colobus + red-tails	49	Red colobus 63.3%, red-tails 36.7%	Red-tails 91.8%
Red colobus + blues	6	Red colobus 100.0%	Blues 100.0%
Red colobus + red-tails + blues	7	Red colobus 57.1%, red-tails 42.9%	Red-tails/blues 100.0%

chimpanzee party arrived. One benefit to red-tails of associating with red colobus, therefore, may be that they receive an early warning of approaching predators—upon which they quickly act.

The other aspect of this association is that, as we have already seen, chimpanzees attack many colobus groups but very few guenon groups. This may be a result of the greater ease of catching colobus, or a preference for colobus meat. Both Gombe and Taï chimpanzees have been reported to catch guenons and discard them, following which they continue to hunt for red or black-and-white colobus (personal observation and Boesch, personal communication). Perhaps red-tails benefit from an association with red colobus in that, if they are surprised by chimpanzees, the colobus rather than they will be the targets of predation attempts.

There is a widespread assumption that grouping has some mutual benefit, but this has seldom been documented. Gautier-Hion and colleagues (1983) showed that two Central African guenon species, *Cercopithecus pogonias* and *C. cephus*, enjoyed mutual rewards for associating. *C. pogonias* alarm called to arboreal predators while *C. cephus* gave alarm vocalizations to terrestrial threats, and each species responded to the other's calls.

Another important issue in studies of mixed-species groups is which species initiates and maintains the association, and why. Several primate field studies have found that mixed-species associations are not mutually maintained. By examining association patterns and noting which species was most often responsible for making and breaking the contact, Cords (1990) showed that red-tailed monkeys in the Kakamega Forest of Kenya maintained contact with blue monkeys. These two guenons utilized each other's presence to reduce surveillance and to spend more time feeding.

At Gombe, red colobus were often members of mixed-species associations, but did not initiate or maintain them. Instead, nearly all contacts between red-tails, blue monkeys, and colobus were the result of one of the guenons approaching the colobus and feeding or resting nearby (424 of 448 contacts or 94.6 percent). Moreover, once red-tails were with colobus they frequently departed or moved some distance away, returning repeatedly during the day. The presence of animals with distinctive hybrid features, such as a blue monkey's larger size and gray-blue color but a red-tail's white nose and red tail, enabled me to recognize members of the group. Mixed-species associations at Gombe appear thus to be the result of an active interest by red-tailed and blue monkeys in associating temporarily with red colobus.

In West African forests the olive colobus (*Procolobus verus*) is found in close association with guenons, especially the diana monkey (*Cercopithecus diana*). The colobus benefits, for it is small bodied and forages in widely dispersed groups that could not make effective use of the many-eyes benefit of social predator protection. John Oates and George Whitesides (1990) suggest that this association is the result of colobus actively following dianas as part of an evolved strategy of olive colobus to seek safety in numbers. Because it has a different, more frugivorous diet, the diana monkey may gain increased predator detection through this association with only slightly greater food competition.

In the neotropics, Boinski (1989) found that Costa Rican cebus monkeys followed squirrel monkeys, maintaining contact with them and perhaps deriving antipredator detection benefits while the squirrel monkeys were essentially neutral in the relationship. Marina Cords (1990) has pointed out that interspecific mutualistic or commensal relationships vary greatly from forest to forest, depending on local resource distribution and predator risk. Colin and Lauren Chapman (1996) found that the feeding and predator-detection advantages for monkeys of the Kibale Forest differed greatly among the various species in mixed-species groups.

An important reason why one species is found in contact with another may be that it is unable to drive the other species away. It may not be desirable for red colobus to associate with red-tailed monkeys, but it also may not be possible or energetically feasible for them to drive away red-tails away when they approach. If not, this relationship would be commensal, with one side obtaining a benefit while the other species does not incur cost or benefit. Alternatively, one species may benefit while the other suffers some loss of feeding efficiency or risk to predation. I have never seen a red colobus attempt to drive a red-tailed or blue monkey, but colobus do not seek out contact with these species. Instead, they appear to be the nuclei around which the two guenon species gather.

A series of recent field experiments have addressed the mixed-species associations of red colobus in another region of Africa. In Taï National Park in Ivory Coast, red colobus associate often with diana monkeys. These species spend 62 percent of their time together, more than would be predicted by chance (Holenweg et al. 1996). This association pattern appears to result from the intense predation pressure faced by Taï red colobus (Noë and Bshary 1997). In Sierra Leone, these species do not associate often in areas of low predation pressure (Whitesides 1989), suggesting that their association is a response to chimpanzee predation.

In Taï, the behavior of red colobus and a guenon are markedly different from Gombe. Taï red colobus actively seek out associations with diana monkeys and respond to taped playbacks of chimpanzee calls by moving closer to the dianas. Taped calls of leopards do not elicit this response from colobus. Moreover, the colobus associate with dianas significantly more often during the hunting peak, and also on days when they have heard chimpanzee calls (Noë and Bshary 1997). This field study is compelling experimental evidence of primate behavioral responses to predators.

Mixed-species groups occur in tropical bird flocks as well as in primates. Moynihan (1962) identified roles in mixed-species neotropical bird flocks as nuclear and attendant. Nuclear species included both active nuclear—those that seek to maintain contact—and passive nuclear—those that are joined. Attendants are species that join and leave the nuclear species. The benefits of mixed-species flocking for the nuclear species are unclear; perhaps there are none. Benefits to other species, at least in bird flocks, appear to be predation related. Gombe red colobus are a nuclear species, whereas Taï red colobus are attendant in certain seasons. Because the members of the colobus group have a genetic stake in protecting and promoting each other's survival, attendants such as red-tails may receive an antipredator benefit at little cost.

It appears therefore that predation influences the behavior of red-tailed and blue monkeys in relation to red colobus. How are food needs affected by this association? I have showed that red colobus feed significantly less often in the presence of red-tails, indicating that there is no distinct benefit to them in socializing with the other species. If red-tails benefit from this relationship and colobus cannot or do not attempt to keep them away, we might ask why red-tails do not associate with colobus on a permanent basis. It may be that the two species have different diets, preventing red-tails from remaining in the same food patches with the colobus throughout the day. If their diets do differ, interspecific competition for food is mitigated, and red-tails probably spend as much time as possible with colobus so long as they can obtain the food they need.

In fact, they may do exactly this. Red colobus eat a diet that is high in new leaves and low in fruit and invertebrates. Based on other field studies, the red-tailed monkey diet is high in fruit (60 percent of the diet in Kakamega Forest, Cords 1987). The amount of time that the two species can spend together and still eat their preferred diets is limited and may dictate the strategy of the red-tails for associating with the colobus. Ecological differences may explain the difference in association patterns between Kibale and Kakamega for blues and red-tails as well. In Kibale, where the diets of the two species are more divergent than at Kakamega, the two spend less time together (Cords 1990). At Taï, however, communal foraging in the same food patches explains only a small amount of the association time between red colobus and diana monkeys (Wachter et al. in press). These results further reinforce the notion that such associations are more about predation than about food.

If the predation risk is intense enough, we might expect that many species of monkey would spend as much time together as their dietary requirements allow. Single species might not form huge groups, owing to intragroup feeding competition, but multiple species with different diets could do so—hence the persistence of mixed groups in some species but not in others. The behavior of red-tails when they are with

Gombe red colobus is distinctive; they tend to arrive, remain with the colobus in a food patch, and then move off for a few hours only to return later. They appear to take advantage of the colobus when it is possible to do so and still obtain the food they need, going elsewhere when the colobus are feeding in undesirable food patches. They may even periodically monitor the colobus feeding patterns to decide whether to forage alone or to join the colobus. Gombe red colobus, meanwhile, appear to be almost oblivious to the presence of their associates. Mixed-species associations may correspond in time and space to feeding patterns that allow them to utilize each other while spending enough time apart to forage effectively for their own preferred foods. The amount of time spent apart should correspond to the dietary niche differences between the two species. Instead of seeing mixed-species groups as a paradox of sociality, we might instead ask why these groups do not form more often or become more long lasting assemblages in areas of high predation risk.

1. Kakombe Valley seen from the rift escarpment, showing the home range of the two red colobus study groups and the core area of the Kasakela chimpanzees.

2. On the upper slopes of the valleys the vegetation is mainly miombo woodland.

3. View looking north showing wet-season foliage (November).

4. Same view, taken during dry season (August).

5. Brothers Frodo (left), age 17 years, and Freud, age 22 years.

6. Goblin, age 28 years.

7. Wilkie (left), age 21 years, and Prof, age 22 years.

8. Evered, estimated age 41 years.

9. Beethoven, estimated age 25 years.

10. Atlas, age 25 years.

11. A party of chimpanzees, led by Frodo (far right), approaches colobus males of J group, 18 July 1992, Kakombe Stream.

12. Frodo stares at a male colobus defending his group a few meters overhead.

13. Frodo rushes at the male colobus, attempting to scatter their ranks and reach an immature.

14. Frodo hand-threatens a male colobus as he attempts to reach the females and their offspring, out of view to the left.

15. On 7 October 1992 Atlas captures a juvenile colobus of J group, using his arms and one foot to pin the prey to the tree limb.

16. Atlas kills the juvenile colobus with a bite to the base of the skull.

17. An adult male Gombe red colobus.

18. A female red colobus of J group carrying an infant on her abdomen.

19. An adult male chimpanzee is watched vigilantly by a male red colobus in KK5 before a hunt begins.

20. After catching a juvenile colobus on 7 October 1992, Atlas is approached by his younger brother Apollo (left) and the immigrant swollen female Trezia.

21. Atlas withholds the colobus carcass (visible hanging from his left hand) while copulating with Trezia.

22. The politics of meat: alpha male Wilkie shares with his ally Prof.

23. Beethoven eats an infant colobus while a swollen Gremlin begs for meat. Gremlin's son Galahad looks on. Note Beethoven's erect penis.

24. Oil palm nuts provide more calories and more saturated fat per gram than monkey meat.

25. Frodo.

Let us now talk in a little more detail about the Struggle for Existence.

—CHARLES DARWIN, *The Origin of Species* (1859)

7. Confrontation

In spite of the antipredator behaviors that Gombe red colobus employ to avoid becoming targets, they are attacked regularly by chimpanzees. The study groups occupying the core hunting area of the Kasakela chimpanzee range were attacked by chimpanzees on average every 80 days for W group (1991–1993) and every 66 days for J group (1991–1995). Most groups, living in areas less frequented by chimpanzees, would be targeted less often, but nonetheless are subject to repeated predatory attempts.

The moment they hear or see chimpanzees approaching, both red-tailed and blue monkeys disappear through the forest canopy. Red colobus rarely attempt to flee; instead, they sit and wait to see if the chimpanzees will decide to hunt. Fleeing is adopted as an antipredator strategy when it is the most effective means to escape predation relative to some other behavior, such as careful vigilance or a mobbing counterattack. While red colobus are skilled leapers, they are not agile at fleeing from a predator through the forest canopy.

In 10 of the 11 occasions when I saw red colobus attempt to scatter and flee en masse from chimpanzees rather than stand and fight, chimpanzees made at least one kill. This success rate is about twice the overall hunting success rate, so that fleeing is probably not an effective way to escape chimpanzee predators. At certain times and places it may be more effective to flee, but in other circumstances that is not the best option.

For instance, flight from chimpanzees may succeed in high-canopy forest where tree crowns dovetail to form a continuous blanket of foliage high above the forest floor. There the best tactic for a monkey trying to escape a predator approaching from below may be to run from one tree crown to the next, using branches too small for the predator to pursue. This strategy is in fact the one employed by red colobus at Taï

(Boesch 1994b). Fleeing may also help the prey in other less obvious ways; the spectacle of many prey animals moving in different directions simultaneously may confuse the predator enough to allow potential victims to escape. This tactic works occasionally at Gombe, as noted by Goodall (1986). But when caught in a small or scattered patch of trees (a common circumstance for Gombe red colobus), trying to get to the next tree crown in hope of escape may be the worst tactic to adopt. Other stratagems may be required.

Alarm Calls

Alarm calling is a paradoxical form of vocal communication. The caller attracts attention to himself or herself and thereby makes it less likely that others will be killed. Since animals are expected to behave selfishly to gain reproductive benefits at the expense of neighbors, the evolution of this apparently altruistic behavior is difficult to understand. The fundamental paradox of alarm calls was first addressed by Paul Sherman (1977), who studied kin-based altruism in calls of wild ground squirrels (*Spermophilus* sp.). He showed that they were given selectively by animals whose kin were nearby and therefore likely to benefit from the alarm.

Like ground squirrels and other social mammals, red colobus vocalize in a wide variety of contexts, and assigning meaning to those calls does not always follow from understanding either the context or the acoustic structure of the call. Most of our information about the alarm calls used by arboreal forest monkeys derives from studies using tape-recorded playbacks of predator calls (Hauser and Wrangham 1990). Cheney and Seyfarth's (1991) work on social knowledge of vervet monkeys is an example of what can be learned in a field setting using playback recordings in controlled situations. The naturalistic study of alarm calls requires one to be present with notebook open and tape recorder running for a large sample of predator-prey encounters. Important questions include whether red colobus have different calls for different predators (as vervets do) and whether red colobus recognize the alarm calls of other species.

When red colobus hear the pant-hoots of approaching chimpanzees, their first response is to reposition themselves in anticipation of an attack. This movement is accompanied and perhaps facilitated by alarm calls from adult members of the group. These initial alarm calls are the chists in Struhsaker's vocalization categories from Kibale red colobus. They are given almost entirely by adult and subadult males, with a small percentage (unmeasured, but probably fewer than 5 percent) given by adult females. Detecting individuals in the act of calling in the canopy overhead is difficult, and I could not distinguish male callers from female callers. Because Gombe females migrate but males apparently do not, and because of the long tenure of males in their group, males are likely to be alarm calling alongside other related males, at infants who are either his own or a codefender's offspring. Furthermore, the risk of mortality during a hunt is substantially lower for males than for females. The combination of reduced risk and the incentive of warning close kin may contribute to the frequent alarm calling of males, whereas females tend to be silent.

If chimpanzees arrive and begin to hunt, other colobus warning calls accompany the chist call (see Table 5.1). The wheet is given by colobus as they retreat higher into tree crowns. When a colobus is grabbed by a chimpanzee, a higher-intensity, louder version of the chist is given both by the victim and by male colobus as they counterattack the chimpanzees. A graph of the frequency of alarm calls per minute per adult male colobus as chimpanzees approach is bell shaped. At a distance of about 75 meters, the male colobus begin to alarm call. Their calls increase in frequency until the chimpanzees are quite close or within sight, at which point the entire colobus group falls silent. With the colobus now in position to defend and counterattack if necessary, the chimpanzees decide whether to undertake the hunt. If they climb into the trees, the chorus of colobus alarms rings out (Stanford 1995a). What follows is chaotic activity and frenzied noise. The noise is the combined alarm calls of the adult colobus, which can be heard at a great distance. It is usually impossible to count the calls by ear because so many animals are calling virtually in unison. Observations at Gombe also suggest that crowned eagles too may home in on colobus prey by hearing the sounds of their alarm calls (Wallis and Msuya, unpublished).

The Meaning of Red Colobus Alarm Calls

Red colobus give alarm calls in response to both arboreal and terrestrial threats—including chimpanzees, birds of prey and humans. All 119 encounters between chimpanzees and red colobus involved long alarm-call bouts, whether or not a hunt ensued. The three close encounters that I saw between eagles (two crowned eagles and one martial eagle) and colobus elicited brief but intense bursts of alarm calling from the colobus, though an eagle or a small raptor such as a black kite (*Milvus migrans*) soaring overhead did not necessarily elicit alarm calling. The approach of humans always elicited alarm calls, although such alarm calling ceased within a few minutes if the observer remained stationary and quiet or if the animals were habituated. The approach of a party of chimpanzees was no more likely to produce alarm calls from J group than was the approach of a group of tourists, if J group was overhead and the chimpanzees or humans walked along the trail beneath ($t = 2.70$, $p = .55$). I made real-time recordings of the encounters between colobus and both chimpanzees and humans, and also fortuitously recorded one of the eagle encounters. There is no obvious difference in the sound of the calls given in response to the three perceived threats, (acoustical studies based on the tapes must still be done). Other field studies (Cheney and Seyfarth 1988, vervet monkeys; Zuberbühler et al. in press, diana monkeys) have showed that monkey alarm calls may identify predators as individual species.

One test of the red colobus' understanding of the calls they hear is whether they respond to the alarm calls of other species. At Gombe they frequently hear the alarm calls of red-tailed monkeys, the *pyow* loud call of blue monkeys, the alarm barks of baboons and bushbuck antelope, and the waa-barks of chimpanzees, all in reaction to some perceived threat nearby.

I examined colobus response to these calls by noting all the alarms I heard given by other species while I was watching J and W colobus groups between 1991 and 1993. The calls of red-tailed and blue monkeys and those of bushbuck usually elicited an alarm response from the colobus as well (Figure 7.1). The results were more ambiguous for responses to the alarm calls of chimpanzees and baboons. For chimpanzees, I noted waa alarm calls, which signal danger such as a python, leopard, or strange chimpanzee. In half of the instances when they heard chimpanzee waa-barks, the colobus also alarm called. This was a higher rate than in response to the pant-hoots that often accompany the approach of chimpanzees (56 of 254 distant pant-hoots were responded to by colobus, or 22 percent).

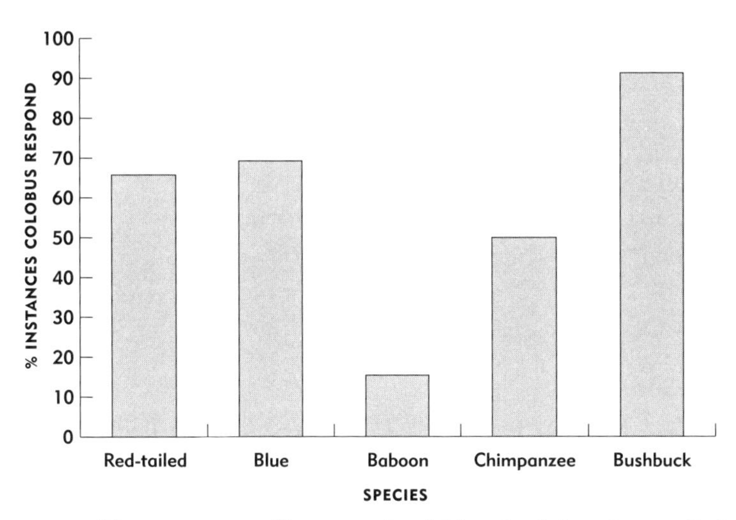

FIGURE 7.1. The percentage of instances in which red colobus alarm called after hearing alarm calls of other species at Gombe. J group, 1991–1994. Samples: red-tailed monkey, 105; blue monkey, 26; baboon, 58; chimpanzee, 8; bushbuck, 23.

Chimpanzees pant-hoot in many contexts throughout the day; red colobus thus appear to be able to distinguish pant-hoots that may signal an approach by chimpanzees from other calls. If colobus alarm call in response to another species, the alarm calls of it may be that they understand the meaning of those calls. However, when a bushbuck alarm barks and a colobus alarm calls immediately afterward, it could mean merely that the colobus was startled by the loud bark that erupted from a nearby thicket and responded with alarm.

The Object of Alarm Calls

Alarm calling has evolved in response to the risk of being attacked. Giving an alarm call when a predator is seen or heard is usually an effective antipredator strategy when there is enough time to react to the predator and influence the odds of successful evasion. If the major predators are species that approach with stealth and strike without warning, the utility of alarm calling may be limited. But if the predator strikes more than once in a predatory episode, other individuals could benefit from calls even after one group member had been attacked or killed. Alarm calling should be useful—and its use favored by natural selection—when it can make a difference in the outcome.

Whether or not alarm calling makes a difference depends on the circumstances and the nature of the predator and of the prey. For example, most of the existing work on primate alarm calls has focused on the initial call given in response to the detection of the predator, as in Cheney and Seyfarth's (1991) work on vervet communication. Amboseli vervets clearly use alarm vocalizations in relation to known and understood relationships and kinships within the group. The calls are given with some voluntary control and may even be selfish acts of deception, as when an animal is diverted from its food patch by a specious alarm call.

The Redundancy of Red Colobus Alarm Calls

A clear and effective first alarm call heard and heeded by other group members probably confers an advantage, but in some circumstances the duration of the calling bout may be equally meaningful. Extended repetitive alarm calling characterizes many primate repertoires, including those of vervets and red colobus monkeys (Struhsaker 1975; Cheney and Seyfarth 1991). This tonic communication (Schleidt 1973) may keep members of the group on their toes, alert to the presence of a predator that has not struck yet but may do so at any moment. This function has been shown for some calls given by California ground squirrels, *Spermophilus beecheyi* (Owings and Hennessey 1984).

The most salient and paradoxical aspect of red colobus alarm calls is their repetitive and seemingly redundant pattern (Table 7.1). These calls are never given merely as single calls to arms. Instead, they begin well before the chimpanzees arrive, continue in almost rhythmic unison throughout the hunt, and cease only after the chimpanzees have stopped hunting (whether successfully or not) and departed the scene. This pattern might seem maladaptive given the function of alarm calls in other primates and most social mammals. In other species the calling is more or less over once everyone present is aware of the danger.

The key difference between colobus alarm calls and those of other species may be the nature of the predators. For an ambush predator like a leopard or eagle, an instantaneous alarm note is essential because the predator strikes quickly and with little notice. Chimpanzee predators, on the other hand, approach slowly and noisily and the colobus' pre-encounter response is spatial repositioning rather than early warning. For red colobus repetitive alarm calling may therefore

TABLE 7.1. *Alarm call persistence: percentage of red colobus–chimpanzee encounters in which red colobus alarm calling continued for varying periods (based on more than one call heard per minute for entire group).*

TIME INTERVAL	% INSTANCES WHEN > 1 CALL/MINUTE IS HEARD
During encounter or hunt	119/119 (100.0%)
Up to 10 minutes following end of encounter	101/119 (84.9%)
Up to 30 minutes following end of encounter	26/119 (21.8%)

be more advantageous than the first call alone. (Note that other primates may also call redundantly during predator encounters, including Cheney and Seyfarth's vervet monkeys.)

These repetitive calls given in predator encounters may be tonic communication, using the additive effect of new signals superimposed on a sequence of already-received signals to serve a warning function. The apparent function is to maintain a state of vigilance in the receiver of the call and to modify the behavior of group members over a prolonged time span.

The function of repetitive calls has been investigated in ground squirrels (genus *Spermophilus*). Owings and Hennessy (1984) found that long bouts of the same call given by California ground squirrels increased the duration of vigilance by the group, though ever-longer sequences of calls did not necessarily continue to extend the vigilant period. These researchers argued that the callers attempt to "persuade" other receivers of the call that danger remains in the area, and that the callers' success could be measured by how soon the receivers' vigilance began to subside.

Because of the energy involved and the risk inherent in calling in the presence of a group of predators, we can assume that male colobus would not generate such extended call bouts unless they serve a function. They may be warnings to the rest of the group to stay alert and vigilant, even after the hunt has continued for an hour and the colobus are exhausted and not as wary as they may have been earlier. Repetitive call bouts given over extended periods, in comparison to occasional single calls, elicited a much higher rate of looking up by other adult colobus than single calls did (two-tailed t-test, $p < .001$).

One age class of colobus is more vulnerable to attack than any other: juveniles between 15 and 24 months, who are not able to rely on their mothers for protection in the way dependent infants can. They are caught much more frequently than their representation in the population would suggest. This stage of life is without question the most dangerous for a Gombe red colobus, from the standpoint of predation-risk. Repetitive calls may serve to warn naive and older infants to stay alert to the risk of continued attack. The most effective procedure may be to call redundantly, even after the threat appears to have diminished or disppeared.

To test whether such is the case, I compared the calling rates and length of calling bouts for groups that included varying numbers of immature colobus, examining the effects of juveniles, older infants, and the two age classes combined. To control for the size of the group, I considered the calling rate as the number of calls given per adult colobus visible in the group per minute, taking the mean rate over the entire length of the encounter. By making real-time recordings of encounters between chimpanzees and colobus from 1991 to 1993, I was able to count the number of calls on tape and compare it with the number of colobus I noted in the group. The result showed that the more immature colobus there are in a group, the greater the rate of alarm calling (Figure 7.2). This effect was stronger for juveniles of known or estimated age 15–24 months than for infants of estimated age 4–14 months ($t = 5.82$, $p < .001$). Moreover, the greater the number of immature colobus in the group, the longer the repetitive call bouts continued, even well after a threat to the group had disappeared (Figure 7.3). This effect did not hold true for an increased number of adults in the group ($t = 1.07$, $p = .6$).

The Dynamics of Counterattack

18 July 1992: A hunt is in progress in Kakombe Valley, in Albizzia rising above the northern slope of Kakombe Stream. Five adult males led by Frodo are in the tree crown with the male colobus, and Frodo is pursuing a mother who carries a newborn infant less than two days old. Frodo charges repeatedly into the midst of the male colobus; they leap onto his back by twos and threes. Through his persistence and willingness to with-

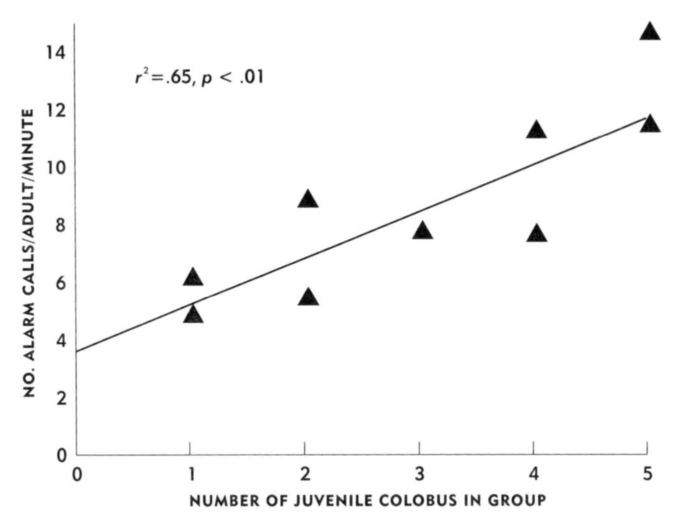

FIGURE 7.2. The number of juvenile colobus in a group that is attacked in relation to the rate of alarm calling by the adult colobus in the group (Gombe, 1991–1995).

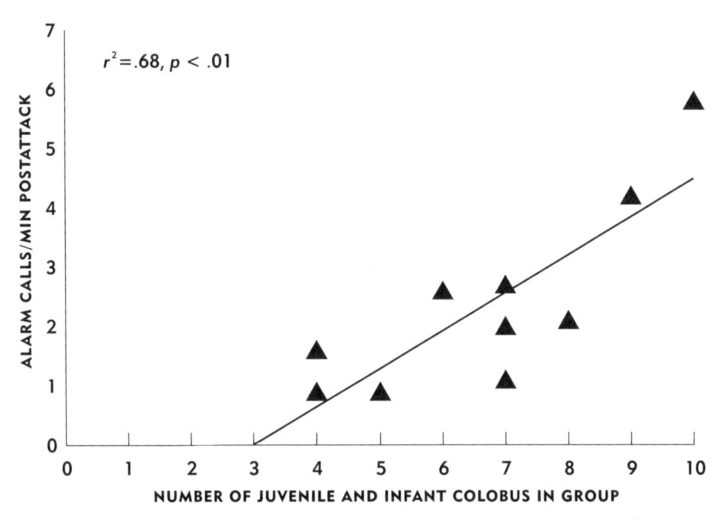

FIGURE 7.3. The number of juvenile and infant colobus in a group that is attacked in relation to the persistence of alarm calling after the attack is over (Gombe, 1991–1995).

stand their bites, Frodo manages to scatter the male colobus long enough to pluck the newborn infant from its mother's abdomen. While the other colobus watch from a few meters away, Frodo eats his quarry. The hunt, which seems to be over, suddenly continues as the mother of the infant races in from the periphery of the scene to attempt to rescue her baby. Even though the infant is now half eaten, she rushes toward Frodo and the carcass. Chased by the other male chimpanzees, she falls from the tree crown to the forest floor, where she is pounced on by juvenile chimpanzees who had been watching the hunt from below. She is killed by these young would-be hunters.

The most dramatic aspect of the relationship between red colobus and chimpanzees is the fearless way male red colobus counterattack, even launching preemptive attacks to drive them away. Mothers sometimes become involved if their babies have been captured, but it is usually only the males who engage in the antipredator aggression that characterizes this species. Counterattacking the predator is assuredly a high-risk behavior. If the predator is large, powerful, and swift and the prey is smaller (the typical case for primate prey), then it may seem an extremely unlikely defense strategy.

Some songbirds respond to the approach of a predator such as a snake or an owl by mounting a group defense known as mobbing. It should be the last line of prevention against predation, because it involves the greatest risk of harm to the mobber. A number of primate species have been observed attacking their predators. Unfortunately, the rarity of observation of such attempts means that there has been little systematic study of counterattack strategies. Many small monkey species mob birds of prey and other predators (Passamani 1995, *Callithrix geoffroyi* mobbing a margay cat, Boinski and Mitchell 1994; *Saimiri* sp.; van Schaik and van Nordwijk 1989, *Cebus* sp. mobbing a snake model; Bartecki and Heymann 1987, *Saguinus fuscicollis* mobbing a snake; Stanford personal observation, *Cercopithecus ascanius* mobbing owls). Chimpanzees are also known to preemptively attack their predators when encountering them in the forest. Christophe Boesch, studying chimpanzees in Taï National Park, reported seeing a party of chimpanzees chase a leopard, which leaped or fell into a hole. The chimpanzees stood around the hole, screaming and throwing sticks and debris at the leopard, before losing interest and leaving. The leopard later escaped (Boesch 1991).

Counterattacking is superficially similar to the mobbing behavior performed by birds. The fundamental components of mobbing behavior among birds are that (1) the species being mobbed is potentially very dangerous, (2) the mobbers pose no serious risk of injury or death to the predator they are mobbing, (3) the mobbers behave collectively, and (4) the mobbers give loud vocalizations while mobbing (Curio 1978). The mobbers in bird species are often quite small compared to the predator and are incapable of doing real harm; thus, in some cases the mobbing takes on the aggressive demeanor of a counterattack, as when many jays attack a hawk or crow. One well-placed jab of the bill can truly injure the larger bird.

In similar fashion, two or three male red colobus weighing 10 kilograms each can seriously injure or even kill a 45-kilogram chimpanzee if a bite is delivered to the right place—or if the chimpanzee is frightened into falling out of a tall tree. Chimpanzees are often wounded on the hands, arms, back and scrotum by colobus during a counterattack. The canines of a male colobus can kill an infant chimpanzee; this may be what limits the participation of female chimpanzees in the hunt. Counterattacking behavior by red colobus is therefore not necessarily the functional equivalent of mobbing, in that its goal may be to injure the predator rather than simply to dissuade it without engagement.

In birds, the proximate impetus for mobbing is twofold. The first counterattacker responds to the imminent threat posed by an approaching predator. This mobber is joined quickly by others who provide the strength in numbers needed to repel the predators. The initiation of mobbing is followed by the decision of groupmates to join the counterdefense or to remain bystanders. Each individual mobber must decide whether to risk joining the mobbing effort. The opportunity presents itself to each male to allow other males to defend against a chimpanzee attacker while refraining from exposing itself to the danger. Whether such males ever attempt to reap the benefits of communal antipredator defense without assuming some of the cost is one of the questions I hoped to answer.

In 119 observed encounters with chimpanzees, only adult and subadult males counterattacked. Immatures do not participate for obvious reasons of size and strength. Their failure to seek protection from their mother leads to frequent capture. Immatures are the preferred prey, but infants must be either taken from their mother's abdomen or grabbed while attempting to return to maternal protection. Juveniles possess neither the haven of a mother nor the savvy required to escape from predators, as evidenced by their peripheral spatial position in the group during encounters (Stanford 1995a). It seems paradoxical that mothers, having invested heavily in an offspring for more than two years, abandon their juveniles. Some females may be pregnant again two years after birth, but their investment of time and energy in the unborn infant is far less than that already embodied in the threatened older offspring. Yet females do little to protect or attempt to rescue juveniles, whereas they often attempt to rescue their dependent offspring who are captured by chimpanzees. They do so even when the risk of death to themselves is great; I saw five females killed by chimpanzees while attempting rescues. Adult females may be constrained from offering protection to juveniles if they have dependent offspring who would certainly be caught if left on their own. Juveniles may also be too large to be carried by the mother during predator encounters. So the mother does not offer protection, nor do juvenile offspring actively seek it.

Female colobus carrying infants are in a vulnerable position in that infants are the favored targets of chimpanzees. But even if females were at greater risk of being caught and therefore did not counterattack with the adult males, their mere presence might inflate the apparent number of mobbers and have deterrent value. Male red-winged blackbirds attempt to exaggerate the number of males already on a territory to discourage new males from entering. This has been called the Beau Geste effect, after the classic film in which besieged soldiers place dead bodies in their fortress garrets to give the appearance of a larger defending force than actually exists (Yasukawa 1981).

During an attack by chimpanzees, the adult males form a phalanx that attempts to stay between the attackers and the females and immatures of the group, and the victims are quite often those who do not stay in relative safety behind these defenders. For adult males, the offspring who are the most frequent targets of the attack may be their own, giving them a genetic stake in protecting the young. Adult females, having transferred into the group, have an investment in at least one offspring but perhaps no more. An adult female may be less likely to risk death in order to protect nonkin.

The Benefits of Counterattack

8 October 1992: In the upper stretches of KK6, W group has been attacked by a party made up of the same chimpanzees who had attacked J group the day before. Buddha, the top-ranking male of W group, whose size and weight approach that of a juvenile chimpanzee, responds to the approach of the chimpanzees along with his smaller groupmates Bones, Black, and Bird. They climb down and abort a climb into the tree by the would-be attackers. Frodo leads the hunting party but is stopped by the male colobus at the first large fork of the trunk of this Albizzia. After a long pause during which all the hunters crane their necks to scan the colobus group above, Frodo appears ready to try his ascent again. But Buddha preemptively leaps down onto Frodo, sending him fear-screaming back down the tree trunk. The other male colobus sit just above, alarm calling. The male chimpanzees linger lower on the tree trunk; in an adjacent tree crown Gimble and his sister Gremlin and her three-year-old son, Galahad, watch the female colobus and their babies intently. Bones and Black launch an attack against this pair, driving Gimble away and nearly knocking Gremlin out of the tree. For a moment Galahad is in danger of being set upon by the male colobus, but he too escapes. As the colobus continue to alarm call from above, the chimpanzees all descend and leave the scene, staring up over their shoulders at the would-be victims.

For counterattacking to have evolved as a red colobus defensive strategy, that strategy must be more successful than other available options. For Gombe red colobus, counterattack lowers the odds of mortality due to chimpanzee predation. Figure 7.4 shows the relationship between the number of male colobus in the group under attack and their success rate in repelling the chimpanzees. There is a weakly positive but statistically nonsignificant relationship between the number of males present and their success. When a colobus group is attacked, not all male colobus responded equally. Comparing only the number of counterattacking males (not including bystanders) with their effect on the hunt's outcome (Figure 7.5), the effect is strong; more counterattackers lead to more successful defenses by the colobus and more failed hunts by the chimpanzees. Male colobus seem to provide protection to their group by jointly defending it. Their behavior thus meets the main criterion for establishing cooperative behavior in social mam-

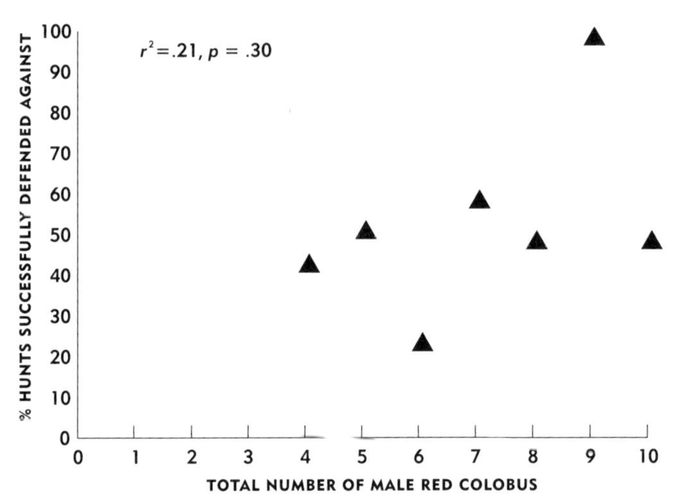

FIGURE 7.4. The effect of the total number of male red colobus in the group on their rate of successful defense against chimpanzee attacks (Gombe, all hunts, 1991–1995).

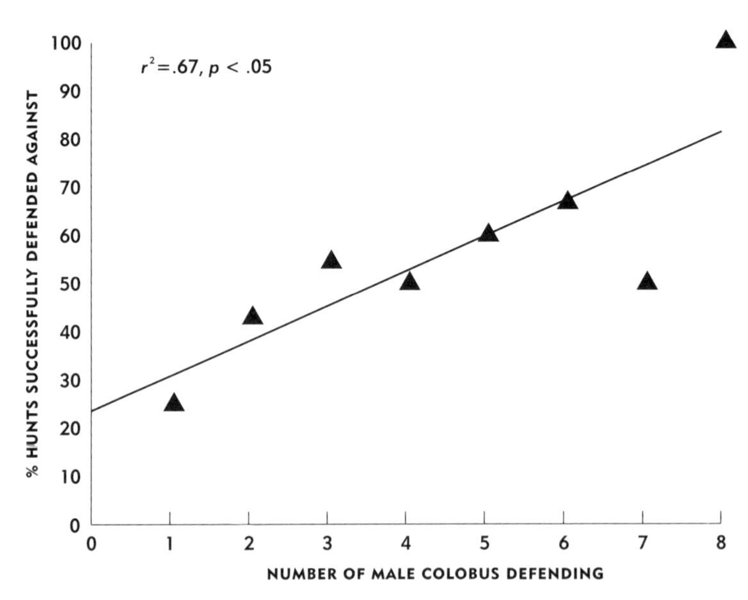

FIGURE 7.5. The effect of the number of male colobus counterattacking on the rate of successful defense against chimpanzee attacks (Gombe, all hunts, 1991–1995).

mals: joint action produces better results than individual action (Packer and Ruttan 1988). We are led to ask how cooperative male colobus really are, if some are joining in the defense and others are not.

The outcome of a counterattack is determined by more factors than just the number of defenders. When we consider the number of attacking chimpanzees as well as the number of defenders, a different picture emerges (Appendix 17). When there are fewer than five male chimpanzees in the hunting party, the number of male colobus counterattacking is the main determinant of successful defense. More than four counterattackers can repel a small number of male chimpanzees, whereas fewer than four cannot. When more than five chimpanzees attack, however, this effect disappears. The joint action of mobbing red colobus is thus effective only when the number of attacking chimpanzees is small. When more than a few male chimpanzees hunt together, no degree of cooperative defense can effectively prevent them from making a kill. Since many red colobus groups have only five or six adult males, they are extremely vulnerable to attack from chimpanzee parties unless most of the males participate in the defense. This finding is supported by many observations of hunts in which the colobus mount a successful defense, only to see it broken because the chimpanzees outnumber the defenders and attack repeatedly until a kill is made.

In nearly every hunt some male colobus join in the defense while others refrain. In 60 hunts an average of 4.2 male colobus counterattacked, while an average of 1.33 did not. The problem of cooperators and cheaters has long been recognized in the study of behavior (Axelrod and Hamilton 1981). What appears to a casual observer to be cooperation is often the exploitation by cheaters of the willingness of others to participate. Some animals exploit others by following as bystanders, potentially obtaining the benefits of group life without assuming the risks. If most of the colobus selfishly exploited the cooperation of other group members, the cooperative defense system would break down.

The evolution of cooperative behavior has been explored using game theoretical models (Packer and Ruttan 1988) in which evolutionarily stable strategies emerge in some contexts but not in others. In the red colobus cooperative defense system, cheaters theoretically should cheat mainly when the number of chimpanzees attacking is large; for the collective efforts of the colobus have the least payoff in such hunts.

Someone is likely to be captured anyway. Only in the case of an attack by a large contingent of male chimpanzees is the best strategy to cheat. At sites such as Taï, where male red colobus cooperation is not effective at driving chimpanzees away (Boesch and Boesch 1989), we should expect to see wide-scale cheating by male colobus in all sizes of hunting party. Whether this occurs is unknown. We might also expect that cheating by Taï colobus is more likely in larger groups, since an attack on 12 males may enable a vigorous defense from some of them while others stay back and avoid the risk of capture.

Boesch (1994c) has suggested that the canopy height in the forest also influences the response by red colobus males to the risk of being hunted. He considered the lower average canopy height at Gombe to dictate the more aggressive behavior of colobus toward chimpanzees relative to Taï. It is a reasonable hypothesis to explain intersite differences in counterattacking behavior, though the most aggressive colobus response seen anywhere is in the Kibale Forest, parts of which contain very tall mature trees. Moreover, Gombe red colobus are most aggressive and most successful at counterattacking in habitats where they can mount an effective defense without being scattered. This habitat is the high-canopy riverine forest bordering streambeds, which most resembles the high-canopy forest at Taï.

The effects of kinship among the male red colobus should change the cheating equation, in that the threshold for cheating should be higher when one shares genes with those who are willing to participate. The mean number of mobbers in all hunts was 4.2 (s.d. ± 1.1). Among these mobbers are probably coalitions that form and act in unison. Such cooperation may occur within the larger context of group cooperation, such as the tendency for Buddha and Bones to be close together during attacks. Coalitions may be based on kinship, but because of the long tenure of males in the study groups in five years I saw only one male reach adulthood and join the ranks of the other counterattackers.

Curio (1978) formulated a series of hypotheses about when and why the prey mob the predator. He reasoned that the adaptive advantages of mobbing behavior must outweigh the loss of time and energy spent in constant vigilance when the risk of being attacked is great. I have inserted additions and revised versions of his theories that apply well to the colobus-chimpanzee dynamic. The following are Curio's nine

TABLE 7.2. *Possible evolutionary explanations for counterattacking behavior by red colobus (adapted from Curio 1978).*

HYPOTHESIS	PREDICTED ADAPTATION FOR MOBBERS	SOURCE
Lethal counterattack	Eliminate risk	Curio 1978
Selfish Herd	Selfishly divide/dilute risk	Hamilton 1971, Curio 1978
Confusion	Reduce predator success rate once hunt has begun	Curio 1978
Move-on	Discourage attack	Curio 1978
Early detection/honest perception	Discourage attack	Curio 1978, Vega-Redondo and Hasson 1993
Cultural transmission	Educate future mobbers	Curio 1978
Keeping vigilant	Keep naïve group members alert to risk	Schleidt 1973, Owings and Hennessy 1984
Alerting kin	Warn kin of impending danger	Sherman 1977
Beau Geste	Dishonest signaling of numerical strength	Adapted from Yasekawa 1981

hypotheses to explain avian mobbing, of which at least six seem to apply to primates as well (Table 7.2).

1. *Lethal counterattack.* Although not the goal of mobbing behavior by songbirds or other small social prey species, lethal counterattack may apply to red colobus, given the weaponry the males have and the frequency of wounds on the bodies of chimpanzees after a hunt. To the extent that it *is* a goal of the male colobus, it distinguishes red colobus antipredator behavior from other forms of mobbing behavior. Vocalizations given by colobus during the hunt may not be comparable to mobbing calls given by birds, because the call to arms may be for help in overwhelming or injuring the predator rather than merely discouraging it.

2. *Selfish herd hypothesis* (Hamilton 1971). There may be numerical safety in mobbing, in that individual odds of capture are reduced. The Selfish Herd effect appears to only weakly influence red colobus grouping patterns, however, since individuals in larger groups are marginally less vulnerable to attack.
3. *Confusion hypothesis.* A joint defense by many individuals may confuse the attacker and cause him to give up. It is one obvious motive for small birds to mob a potential predator; it may also operate for red colobus, but seems to have less influence. I have no empirical evidence supporting or refuting the possibility that male colobus jointly attack chimpanzees in order to confuse them enough to discourage them from hunting. On the other hand, Goodall (1986) interpreted the Gombe chimpanzees' selfish hunting strategy as multiple individual actions aimed at confusing the colobus. It is difficult to infer the degree or importance of confusion, and when a hunt fails or succeeds other more testable hypotheses are available.
4. *Move-on hypothesis.* Mobbing the predator may persuade the attackers that their time would be spent more profitably hunting elsewhere. The predators should leave an area sooner, the longer and more intensely they are mobbed. By preventing the hunt from occurring, the prey may also deprive the hunters of knowledge that would help them in their future attempts to hunt this group.

Red colobus seem to counterattack in order to encourage the chimpanzees to move on. Although colobus groups are not stationary, they travel in the range used by many chimpanzee parties composed of the same individuals who will hunt them repeatedly over months and years. It is conceivable that an expert hunter such as Frodo would be able to recognize individual male colobus and learn their defensive tactics during an attack, obtaining information that he could use to hunt more effectively in the future. However, in the two or three hunts made on the same colobus group by the same chimpanzees over a short time span, the difference in outcome appeared to be influenced mainly by spatial positioning of the colobus and the habitat in which they were encountered.

5. *Early detection/honest perception hypothesis* (Curio 1978; Vega-Redondo and Hasson 1993). The mobbing calls and the mobbing itself may be advertisements to the predators that they have been detected and have lost the element of surprise. Natural selection would be expected to communication systems that have efficient warning signals, such as the colors and sounds that have evolved in many poisonous animals. Potential prey often signal their awareness to would-be attackers, causing the predators to give up the hunt (Kummer 1971).

   Are red colobus signaling honestly to chimpanzees that an attack today would be futile? Because chimpanzees often rely on breaking up the defensive formation, male colobus may be signaling that they are not about to be scattered. We should then expect the male colobus to alarm call most intensely or longest when the number of attackers is few, since this is when the odds are in their favor. In fact, the opposite is the case. A significant positive correlation existed between the rate of alarm calling and the number of chimpanzees attacking ($r^2 = .64$, $p < .001$). This result was probably due to the number of hunters attacking from all sides, which caused a more frenzied response by the colobus. The colobus might also attempt to enhance their numerical strength falsely in order to discourage the chimpanzees (the Beau Geste effect; see below). That neither the honest advertisement nor the false advertisement prediction is supported implies that the colobus do not engage in strategic behaviors designed to prevent an attack by chimpanzees before it has begun.

6. *Cultural transmission hypothesis.* Prey animals may mob a predator in order to learn about his behavior and to teach younger and less knowledgeable group members how to prepare for future encounters. Is counterattacking a learning experience for young and artless red colobus? Curio hypothesized that mobbing by songbirds was designed to teach naive individuals to be afraid of the predator species being mobbed.

   The failure of juvenile red colobus to respond to the threat of imminent death at the hands of the chimpanzees (Stanford 1995a) suggests a strong learned component in avoiding predation. Many studies have shown that alarm calls and perhaps mobbing behavior itself become refined over the course of an immature monkey's development (Cheney and Seyfarth 1991). Curio also suggests that mobbing may simply enhance the learning experience of the immatures in order to avoid future attacks. However, the risk of predation to immature red colobus is so great that a learning experience involving high mortality rates would not seem likely to be an evolved training regimen. The risk inherent in the lesson must not outweigh its survival value. In those rare instances when red colobus males go out of their way to attack a chimpanzee preemptively and drive him off even when there is no hint of an impending hunt, they might be doing so to teach young colobus that chimpanzees are a threat and that attacking them is the best way to reduce that threat.

7. *Keeping vigilant hypothesis* (Schleidt 1973; Owings and Hennessy 1984). Mobbing may occur as a loud and clear warning to other group members, especially kin, that imminent danger exists. Redundant and repetitive vocal signals keep group members vigilant during hunts, making this hypothesis a likely influence in hunts where many immature colobus are present.
8. *Alerting kin hypothesis* (Sherman 1977). The principle of kin-selected behavior is a guiding force for much of the study of the social behavior of animals. As we have seen, the phenomenon that male red colobus cooperatively counterattack while females do not is best explained by genetic benefits, or perhaps the lack of genetic conflicts, inherent in the coordinated behavior of relatives.
9. *Beau Geste hypothesis* (Yasukawa 1981). The Beau Geste effect might be expected to influence female and immature red colobus to participate, even marginally, in mobbing chimpanzees to discourage them from attacking, but I did not see it in Gombe red colobus.

After the Hunt

The aftermath of a hunt is almost as interesting as the hunt itself. Does an attack change the way a group travels or the way members of the group associate with one another in the hours and days following? Carel van Schaik and Tatang Mitrasetia (1990) presented a model of a python to long-tailed macaques in Gunung Leuser National Park, Sumatra, to see if they behaved differently after an encounter in which the faux-python was presumably mistaken for a real one. They found that the macaques behaved and ranged exactly as they had before the encounter and failed to avoid the area of the apparent attack. However, they become more cohesive as a group (though the effect was weak).

I expected that with a fresh memory of being attacked, colobus might avoid certain areas or perhaps certain tree species in which the attack had occurred. Immatures, terrorized by their brush with predation, might spend more time near the males in succeeding hours and days. The difficulty in getting reliable information on these possibilities was twofold. First, after a long hunt the group was often scattered, so that it took hours, or even until the next day, before it was possible to find the animals to observe their behavior. Second, immediately after an attack the group was obviously distressed and sometimes remained that way for many hours; even my own presence disturbed them, no matter how unobtrusive I tried to be. The quantity and quality of information I gathered on the aftermath of an attack was therefore limited, but nonetheless intriguing.

In the minutes following the end of the hunt, the male colobus continue to alarm call repeatedly and the females continue to carry their infants while scanning the ground assiduously. Most group members usually are not visible, having dispersed into other tree crowns or slowly departed as the hunt moved from tree to tree. If the chimpanzees succeed in scattering the male defenses, the subsequent degree of scatter is even greater. However, in a comparison of the mean distance between nearest neighbors in J group one hour after the end of a hunt with the overall mean for days on which no hunt occurred, the males tended to stay more tightly clustered (two-tailed t test; $t = 6.87$, $p < .01$) and the females stayed in closer proximity to males ($t = 5.43$, $p < .01$). Juveniles, who do not benefit from the proximity of group members during a hunt, do not change their association pattern after the hunt ($t = 1.02$, $p < .05$). The overall level of aggression within a colobus group is markedly greater in the three hours after a hunt than in a nonhunt three-hour behavior sample ($t = 6.05$, $p < .001$), and the amount of time spent grooming is lower for all age and sex classes.

Surprisingly, red colobus do not respond to a chimpanzee attack by avoiding that area on subsequent days. When hunts on the same colobus group occurred on consecutive days (usually involving many of the same chimpanzees), it was often in the same spot and sometimes even in the same tree. The distance traveled by the colobus group was significantly greater, on the day following a hunt than on the day of the hunt itself ($t = 8.33$, $p < .001$); it was not different from the day preceding the hunt. The amount of time spent feeding was greater on days after a hunt than on hunt days ($t = 4.12$, $p = .05$), a finding that can be interpreted in two ways. The hunt may be so time-consuming that the colobus lose an hour or more of travel time that day. Alternatively, on the day following a hunt the colobus may compensate for lost foraging time by eating more and traveling farther to find preferred foods. If the latter is the case, then some of the hidden cost of predation risk is revealed: prey species not only lose the time spent in antipredator vigilance and in mobbing, but also lose important foraging and feeding time.

8. The Impact of Predation

Four main factors influence the size and structure of any animal population: birth, immigration, death, and emigration. Their dynamic relationship shapes a breeding population over many generations. In addition, other aspects of the life history of each animal may be under natural selection. Individual female reproductive success, for example, has three main components: birth rate, number of offspring surviving to sexual maturity, and reproductive life span. In turn, the age at first reproduction influences both the reproductive life span and the mean number of surviving offspring. The interval between consecutive births also influences strongly a female's lifetime reproductive output. Natural selection, by favoring changes in any combination of these factors in response to local ecological variables, shapes the reproductive biology of the females of future generations and thereby changes the demographic profile of the population.

The ecology interacts with the life history variables of each animal, and the effect of this interaction is better or poorer lifetime reproductive performance. Only the longest-running field studies can obtain data on wild primate demographic trends because of the problems involved in maintaining continuous recognition of individuals and their genealogies over many years, combined with the difficulty of sustaining and staffing a field study over this period. Observational data on age cohorts for more than two decades exist for Gombe and Mahale chimpanzees (Goodall 1986; Nishida 1990), mountain gorillas (Robbins 1995), howlers on Barro Colorado Island (Milton 1982) and at La Pacifica, Costa Rica (Glander 1980, 1992), and for baboons in Amboseli National Park (Altmann 1980). Similar data are available for semifree-ranging (human-provisioned) rhesus (Sade et al. 1976) and Japanese macaques (Masui et al. 1975; Melnick and Pearl 1987). "Long-term" here is relative; more than 30 years of observation were required at Gombe to record the first fourth-generation birth, while a much shorter time would be needed for monkeys. The precise combinations of extrinsic influences from the physical environment and intrinsic influences from the social milieu are rarely evident. In this chapter I describe aspects of Gombe life histories, focusing on those traits that appear to be influenced by chimpanzee predation. The relatively short length of the study limits the conclusions, but I have uncovered suggestive evidence of the importance of predation as an agent of natural selection. By comparing Gombe with other red colobus sites we can examine potential predation-related adaptations of red colobus reproductive ecology.

Reproductive Ecology

Are red colobus antipredator behavioral adaptations accompanied by reproductive adaptations? An animal population that suffers most of its mortality in a particular season may exhibit individual adaptations to minimize these losses, such as a modification of the birth season. For primates who live in a highly seasonal environment, food may be unavailable in a particular season (for instance, among high-altitude monkey populations in the Himalayas) and nutritional deficits may result (Curtin 1975). Giving birth at a time when food is readily available to the mother may be essential for infant survival in such seasonal habitats. Alternatively, the birth season may be timed for the infant to become independent of its mother and to begin to forage on its own in a food-abundant season.

Birth seasons can also be timed to reduce the risk of predation to the infants. In an extreme example, nearly all females in a Costa Rican squirrel monkey population give birth within days of each other (Boinski 1987). Local predators thus may be swamped with more newborns than can be eaten at one time. Birth clumping probably evolved when natural selection favored females who birthed infants near the middle of the interval during which other females were giving birth. Infants born in close temporal proximity to other infants are less likely to be eaten by the same predator—especially by solitary, territorial predatory species that kill one prey item at a time.

For many primate species, including most colobine monkeys, births take place in every month of the year without significant seasonal variation. Often a less distinct but statistically observable birth peak occurs (Stanford 1991). The timing of a peak may be the product of natural selection that favors females who produce their offspring at particularly advantageous or minimally dangerous times of year. Just as seasonal patterns of birthing evolve, births can also be timed to occur aseasonally to enhance survival chances for the infant. A group of social predators such as chimpanzees have the capacity to kill and consume many infant monkeys in each hunt. Hunting efficiency may even be enhanced since an entire colobus group could become a prime target for as long as large numbers of infants are present. In this case, spacing births uniformly across the year would be advantageous; mothers would be breeding at times when other females are not.

Gombe red colobus are not seasonal breeders. Although there is a slight birth peak in June, there is no evidence of reproductive seasonality based on births recorded in the five study groups (Figure 8.1). Nor is there birth seasonality among red colobus at Kibale (Struhsaker 1975) or at Taï (Galat-Luong 1983), even though both are also preyed upon heavily by chimpanzees. By contrast, two well-studied populations of red colobus who do not experience predation by chimpanzees show significant reproductive seasonality. Western red colobus (*C. b.*

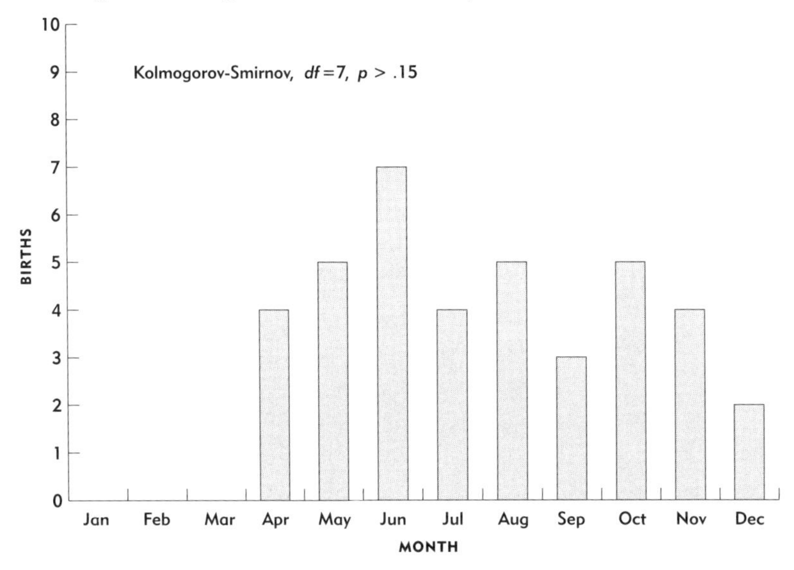

FIGURE 8.1. Monthly variation in births recorded for the five study groups at Gombe, 1991–1995. Incomplete data for January through March.

temmincki) in Abuko Nature Reserve, Gambia, show a strong seasonal pattern of mating and births (Starin 1991), as do red colobus of the Tana River Reserve, Kenya (*C. b. rufomitratus;* Marsh 1981). These populations are separated from Gombe and Kibale by 4,000 kilometers and different habitats but share at least one aspect of their ecology: they live in sites free of chimpanzees.

Gestation length in mammals, including primates, has been a subject of great interest for many decades. What does it mean evolutionarily when one species takes 7 months to gestate an infant while a related species takes only 6? There may be a evolutionary explanation, or the differences may be merely the by-product of developmental exigencies. For most colobine monkeys the average gestation period is 6.0 to 6.5 months, several weeks longer than for the related cercopithecine species (Harley 1985; Struhsaker and Leland 1987). The cause is not body size differences, since some small-bodied colobines gestate longer than larger-bodied cercopithecines. Nor does it appear to be environmental factors such as seasonality. It may be advantageous for a colobine female to produce a more fully developed infant, which necessitates a longer gestation (Boinski 1987; Mitani and Watts 1996).

Red colobus provide an intriguing exception to the longer gestation period of colobines. Gestation lengths for red colobus have been reported from 5 to 6 months (Struhsaker 1975). A small sample of gestation lengths could be reliably estimated among the five Gombe groups by calculating from the frequent mating bouts observed. The average gestation length was only 159 days, about 5.3 months (Table 8.1; $N = 4$). This period is substantially shorter than for other African colobines such the black-and-white colobus (Oates 1974), and dramatically shorter than for some Asian colobines. The Hanuman langur *(Presbytis entellus)*, for example, has a gestation period of about 200 days (Harley 1985).

A short gestation has two implications. First, infants may be delivered when they are slightly less mature than in a species in which pregnancy lasts a month longer. Maternal investment of time and energy is therefore lower in the prenatal period. Second, a female whose infant dies is able to produce another within a shorter time. She invests less time and energy in each infant she produces, so if the infant is lost to a predator in the early weeks or months of life she has mitigated the loss of her investment. A month-long shortening of each gestation period in

TABLE 8.1. *Characteristics of red colobus reproduction in Gombe (1991–1995), Kibale, and Abuko. Some figures are estimates based on age extrapolations of females before the studies began.*

PARAMETER	GOMBE (J,W GROUPS)	KIBALE (CW GROUP)	ABUKO (FOCAL GROUP)
Age at first reproduction	36–48 months ($N = 3$)	38–46 months	50 months ($N = 4$, range = 33–61)
Gestation period	5.3 months ($N = 4$, range = 150– 170 days)	?	?
Interbirth interval (after offspring surviving to 24 months)	24.3 months	27.5 months	29.4 months
Interbirth interval (including after offspring death)	19.2 months ($N = 8$)	24.4 months	28–32 months
Annual birthrate per adult female	0.54	0.49	0.24
Offspring mortality up to 24 months old	64.7%	32.2%	21.4%
Mean offspring survivorship to maturity	21.6%	39.0%	17.9%
Seasonal births	No	No	Yes
Percentage infant mortality due to predation	At least 53.7%	?	At least 40.0%

Sources: Gombe—this study and Watt, unpublished; Kibale—Struhsaker and Pope 1991; Abuko—Starin 1991.

a female's reproductive life span could allow her to have one additional birth during her lifetime.

An additional component of female fecundity is the interbirth interval (IBI), which may also be adjusted in response to local ecological conditions. The interbirth interval, the period between successive live births, varies widely among primate taxa and among individual females. The health and nutritional status of the female can change the interbirth interval. A shorter IBI yields a higher reproductive rate for the female; even a slight increase may be enough to offset the losses to predation from her anticipated lifetime reproductive output. Other field studies have shown that females with shorter IBIs enjoy high fitness (Sommer et al. 1992). In an environment of frequent predation, we would expect infants to be produced more rapidly than at sites where predation is less of a factor. Natural selection is not apt to favor females who invest time and energy in infants likely to be lost to predators. At Gombe, infant mortality due to predation is at least 54 percent (Table 8.1), and possibly much higher. Under such circumstances, shortening the gestation time may be a female's most effective way to increase infant births.

The mean interbirth interval among females at Gombe is 24.3 months following the birth of an infant who survives at least two years. This interval is shorter than reported at both Kibale (27.5 months, Struhsaker and Pope 1991) and Abuko (29.4 months, Starin 1991). The IBI is shortened by the death of a female's previous infant in the early months of life. A mother who has lost her infant begins to cycle again and, since Gombe red colobus are not seasonal breeders, she does not have to wait until the next breeding season to conceive again. Instead, she may resume cycling again as little as 2 months after infant loss and can give birth again following a 5-month gestation. The loss to the mother's reproductive life span would therefore be mitigated, particularly if the infant lost to a predator was very young. Losing a neonate might mean a reproductive setback of as little as 7 or 8 months if a female conceives again quickly. In contrast, a female in the seasonally breeding red colobus population at Abuko who loses her neonate just after the breeding season must wait until the following year to breed again. The lost time plus the second gestation makes a difference of 8 months, a more substantial portion of the female's reproductive life. So little is known about female longevity in the wild that this topic cannot be fully addressed.

The interbirth interval of Gombe females who lost their babies to predators was about 18 months, and for mothers of all disappeared infants, 19.2 months (Table 8.1). This period is approximately 25 percent shorter than among Kibale red colobus and nearly 40 percent shorter than at Abuko. For example, female JN's infant male offspring J9, born at Gombe in June 1994, was eaten by Frodo on 1 September of that year. By July 1995 JN had another infant, again a son (J10), estimated to have been born in June 1995 (he was a neonate when I first saw him in July 1995). The interbirth interval was between 9 and 10 months. Thus JN resumed cycling within 4 months of the death of J9. The brevity of this interbirth interval, combined with the tendency of Gombe chimpanzees to kill mainly young red colobus, has a profound effect on the ability of the red colobus to replace themselves without sustaining severe losses to the population.

The age at which a female begins to reproduce is another aspect of fitness that can increase her reproductive life span. Inherent in female age at first reproduction is a trade-off between using energy from food resources for continued growth versus using that energy for the energy-expensive reproduction that occupies most of her adult life span. When first reproduction is later, females may have a larger body size and perhaps better nutritional status than younger mothers do (Harvey and Clutton-Brock 1985). The age at first reproduction at Gombe is earlier than at either Kibale or Abuko, though the sample sizes are very small (Table 8.1). I observed only two females from birth to motherhood; two others reached sexual maturity and began copulating during my study.

Gestation length combined with interbirth interval produces birth rate—the number of live infants produced per adult female per year. Birth rate, reproductive lifespan, and survivorship of a female's offspring to maturity combine to produce her lifetime fitness. At Gombe the annual birthrate per female is 0.54, or slightly more than one infant produced per female in alternate years. At Abuko the annual birthrate is only 0.24 per female per year while at Kibale it is 0.49. These figures provide further evidence that the fecundity of female red colobus at Gombe is quite high, owing to their short interbirth interval and short gestation. Gombe red colobus reproductive parameters may have evolved to counter the effects of severe predation pressure. To test this hypothesis further, we need to examine the mortality pattern of colobus over time in relation to predation.

Age-Specific Mortality

Gombe red colobus appear to suffer higher mortality rates from predation than most other wild primate populations. Or do they? The predation-related mortality rate for Gombe red colobus up to the age of sexual maturity is about 60 percent. The same figure among Abuko red colobus is at least 40 percent—from a variety of predators, but not chimpanzees. Table 8.2 shows demographic variables of age-specific mortality of Gombe red colobus based on survivorship from birth to adulthood.

Mortality is about 30 percent during the first 6 months of life, which is similar to the mortality of many other wild primate populations. Mortality from observed predation is about 18 percent during this period. Mortality rates then fall during the postneonatal stages from 6 to 12 months, after which they increase sharply. The mortality rate for juveniles 18 to 24 months old is 28 percent, including 16 percent due to chimpanzee predation. This period is very dangerous for maturing colobus. If a juvenile survives to age 2.5 years, its chances of reaching adulthood are excellent, as mortality drops to much lower levels for the subsequent two years of preadult life.

By the time both males and females have reached sexual maturity and females are having their first infants, only about 20 percent of their birth cohort has survived. This rate is approximately the same as for other primate populations in which mortality has been measured; Amboseli vervets, for which predation is the major risk factor (Cheney et al. 1988), Barro Colorado howlers (Froelich et al. 1981), sifakas in Ranomafana National Park, Madagascar (Wright 1995), and toque macaques at Polonnaruwa, Sri Lanka (Dittus 1975).

Figure 8.2 shows that the survivorship of maturing red colobus is very low in early life stages, leveling off near maturity and apparently remaining fairly stable through adulthood, given the low rate of death by predation for adults. This pattern probably results from the tendency of chimpanzees to catch immature colobus. We do not know whether these population parameters would be very different at, say, Taï, where adults are most often taken. The casual observer's impression that the Gombe red colobus are under severe population stress and may face extinction at current predation rates is therefore almost certainly unfounded.

After a history of neglect, the juvenile stage of the life cycle has recently been the focus of much interest (Perreira and Fairbanks 1993). More than just a transition through which every animal must pass between maternal dependency and full independence, the juvenile life stage is itself subject to natural selection. They may be selected to

TABLE 8.2. *Age-specific mortality of immature Gombe red colobus.*

AGE INTERVAL (MONTHS)	ENTERING AGE INTERVAL (n_x)	DYING (d_x)	DEATHS BY PREDATION (d_{px})	MORTALITY (q_x)	OBSERVED PREDATION MORTALITY (q_{px})	SURVIVORSHIP TO AGE 4 (l_4)
0	51	16	9	0.313	0.176	1.000
6	35	6	3	.171	.086	0.687
12	29	4	2	.138	.069	.569
18	25	7	4	.280	.160	.490
24	18	5	2	.278	.111	.353
30	13	1	1	.077	.077	.255
36	12	1	1	.000	.000	.235
42	11	0	0	.100	.100	.216
48	11	1	0	.111	.000	.216
54	10	—	—	—	—	.196

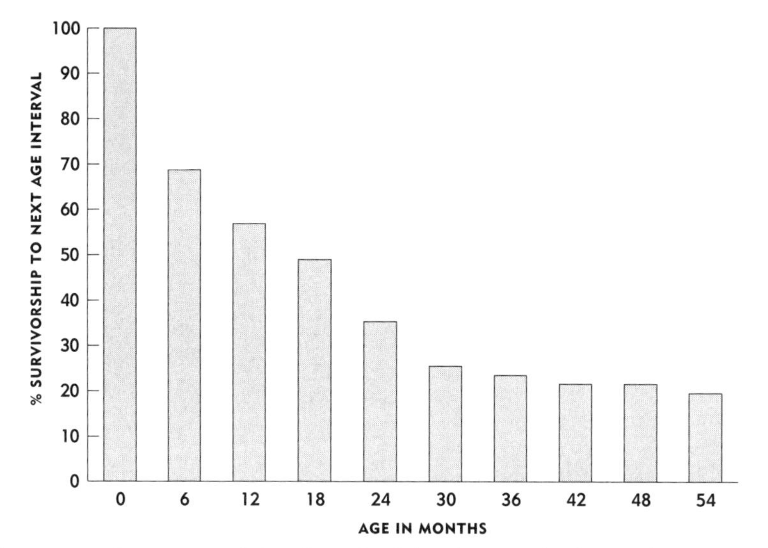

FIGURE 8.2. Percentage of Gombe red colobus born in the five study groups at Gombe, 1991–1995, surviving to the next age interval.

develop more quickly under certain ecological conditions (Pagel and Harvey 1993). Juveniles may reach maturity sooner to avoid the risk of dying before leaving any progeny. This would be particularly true of species in which mortality risk is a function of body size (Pagel and Harvey 1993), and might apply to immature colobus who are far more vulnerable to predators than adults are.

Physiological constraints on the age of first reproduction exist, however, as well as lower limits on body size. One would expect juveniles to engage in less risky behavior than adults—and to some extent they may do so (Janson and van Schaik 1993), even at the cost of reduced food intake (Janson 1990). Yet there is also evidence that independent immatures frequently find themselves in the most exposed and dangerous locations within a social group (Cheney et al. 1988). Adult females who have both a juvenile offspring and a neonatal infant or fetus must decide which to protect in the event of a predatory attack. Such protection in turn must be balanced against the future reproductive value of the female's own life.

The costs and benefits of protecting one's groupmates may be weighed by a risk assessment model (Clark 1994), in which one would think that the age class with the smallest maternal investment and the lowest reproductive value at the time of the attack would be neglected in order to protect older offspring. Red colobus mothers, as we have seen, do the opposite by protecting their younger offspring. They may be unable to provide the protection that juveniles would require, were they to rely on their mothers for protection from chimpanzees.

Lives of Female Monkeys

On 10 November 1992, several of us are following a party of the Kasakela chimpanzees through Mkenke Valley. At 1300 the party, which contains seven adult males (Wilkie, Atlas, Frodo, Freud, Prof, Beethoven, and Tubi), hear colobus calls and race toward them. When I catch up they have arrived at a stand of (Pseudospondias microcarpa) *in which sits a large group of colobus, with many females and their infants. Instead of hunting, the chimpanzees climb into the adjacent mguiza and give excited food calls. They begin to feast on the small, ripe, yellow-green fruits. The colobus, meanwhile, alternate feeding with giving chist alarm calls.*

Thirty minutes later, the mguiza feeding bout ends and the chimpanzees turn their attention to the colobus in the next tree. The hunt begins, and after 20 minutes of unsuccessful rushes by the male chimpanzees and counterattacks by the male colobus, Prof breaks through the colobus ranks and attacks a female with a tiny infant clinging to her chest. The female turns and runs to the other end of the tree crown with Prof in pursuit. She leaps and runs through two more tree crowns and finally stops, her sides heaving with exhaustion. Prof approaches and sits down next to her, directly over my head. The female colobus is far from the protection of the males, but she makes no attempt to flee or attack. Prof slowly and deliberately, almost gingerly, extends his hand toward her chest and plucks the infant from her. She watches silently as he puts the baby in his mouth, kills it with a cranial bite, and eats it a meter away from the unharmed mother.

Females of many social animal species are vulnerable to being eaten by predators when away from the safety of their groups. In the case of female red colobus, the danger may be enhanced because the females migrate between groups. Each adult female and her offspring therefore live in a group with many unrelated females. We have already seen that they do not cooperate to defend themselves during predatory encounters, and that protection is provided by coalitions of adult males. While migrating, a female is alone and faces predators without support from groupmates.

Our understanding of the behavioral strategies of female monkeys has lagged the study of males through most of the history of primate field study. Certainly it is logistically difficult to follow females who have left their natal groups. Furthermore, we know very few female red colobus life histories, including how and why they make decisions related to their lifetime reproductive goals. Migration is an example. In most cases (Struhsaker 1975) immigrants are observed entering a group, whereas the fate of departed females is unknown. Nearly every primate higher taxon includes both male-philopatric and female-philopatric species, and in some taxa both sexes transfer (Moore 1984). The chimpanzee is one of the few strongly male-philopatric species; coincidentally, the red colobus is another.

The most detailed study of red colobus dispersal and migration patterns was conducted between 1978 and 1983 by E. Dawn Starin in the Abuko Nature Reserve. In this tiny riverine forest patch, red colobus (*C. b. temminckii*) occur in high density. Females typically leave their group before reaching sexual maturity. Males tend to remain in the natal group, though they often leave during adolescence and wander alone for a time before returning. In contrast, females usually settle in a new group, tending to choose one to which other females from their former group have also transferred (Starin 1991). This pattern of transfer does not appear in other studies. In research on the endangered Tana River red colobus *(C.b. rufomitratus),* many females and two males migrated to or from the study group; there were many extragroup males as well (Marsh 1979). In Kibale National Park, all migration of adults was by females, although males sometimes left and later returned to the natal group to settle (Struhsaker 1975).

At least seven females appeared in the five Gombe study groups between 1991 and 1995: two in J group, two in MK group, one in C group, and two in AK group (Table 8.3). The rate of immigration into the five groups was 0.28 individual per group per year. All were adult or subadult females; their place of origin was unknown in all cases.

The circumstances of arrival were seen in only one case, that of an adult female who appeared in J group in late June 1992. She gave birth to an infant only 3 weeks later. In spite of her pregnancy she was seen copulating with at least two of the J-group males (JS, JK) in the week following her arrival. I had seen the same pattern in a previous field study: a female capped langur *(Presbytis pileata)* immigrated to the study group and mated frequently with the one resident male during the initial postarrival period (Stanford 1991). Given her late-stage pregnancy at the time, the pregnant immigrant colobus was probably

TABLE 8.3. *Characteristics of migrant females at Gombe, 1991–1995.*

FEMALE	AGE CLASS	PARITY	NEW GROUP/ DATE OF IMMIGRATION
J2	Adult	Multiparous (gave birth 6/92 after immigrating)	J, June 1992
J3	Adult	Mulitparous	J, 1993
M8	Subadult	Nulliparous	MK, 1992
M10	Subadult	Nulliparous	MK, 1994
C11	Subadult	Nulliparous	C, 1993
A9	Adult	Multiparous	AK, 1993
A12	Subadult	Nulliparous	AK, 1994

engaging in socially strategic behavior, aimed at easing her acceptance into the new group, rather than reproductive behavior. This female's newborn infant was killed by chimpanzees on the first day of its life. Unfortunately I had to leave Gombe in early August of 1992 and never learned when the immigrant female began to cycle again or to copulate with group males.

The other six immigrant females arrived without infants and all gave birth later, at a date by which one of the group males could have been the father. This relative stability is a feature of Gombe red colobus groups to which I will return later. It is worthwhile to consider why female colobus remain in groups that are being attacked regularly by chimpanzees rather than transferring to other groups in less predation prone parts of Gombe.

Stability is a feature of male reproduction as well. The reproductive careers of males are understood even less than those of females, because of the lack of visible signs of reproduction. Even mating success is not always an accurate predictor of reproductive success in male monkeys (Strum 1982; Smuts 1985). The tenure of male red colobus at Gombe obviously was long and stable; male membership in the five study groups changed very little over five years. Only two changes in the males occurred in J group: one adult male disappeared while another who had been a subadult in 1991 reached adulthood.

The Fate of the Study Groups

J group, in spite of occupying the core area of the Kasakela chimpanzees and suffering repeated attacks from them, did not change markedly in size or age/sex structure from 1991 to 1995. This is true despite the number of group members killed (Appendix 18). At the end of 1991, J group had 24 individuals and an adult sex ratio of 0.54. In 1992, during which at least 9 members of J group were eaten by chimpanzees and one other individual disappeared, group size was 20 and the sex ratio was 0.58. Group size then increased, and the sex ratio approached 0.50 for the next four years, so that by mid-1995 the group was back to 26 individuals.

Sex-ratio changes in all five groups are presented in Figure 8.3. During this period the ratio of immature to adult animals also varied little (Figure 8.4). The loss of many immatures to predation was offset by births and by the immigration of two adult females. In July 1992, J

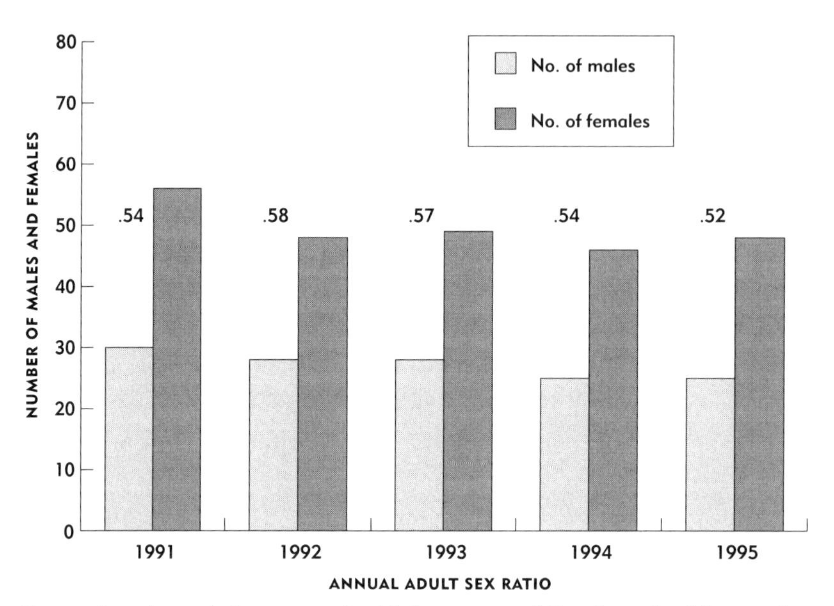

FIGURE 8.3. Annual changes in the adult sex ratio of Gombe red colobus in the five study groups at Gombe, 1991–1995.

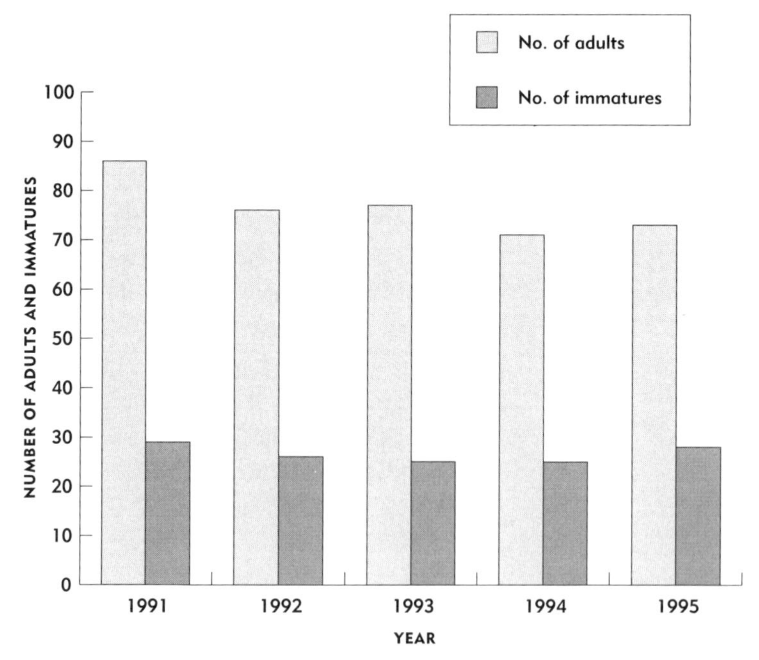

FIGURE 8.4. Annual changes in the ratio of immature to adult Gombe red colobus in the five study groups at Gombe, 1991–1995.

group suffered a predatory attack in which a female (JV) and her newborn infant were killed by Wilkie and Frodo. Three months later, on 7 October, a party of 32 chimpanzees attacked J group and killed 7 individuals, reducing the group by nearly one-third in 90 minutes. That night a new infant was born and J group began to increase again.

Perhaps because it had been depleted of its immatures, J group was not a frequent target of the chimpanzees during the next three years, and its size increased to 26 individuals. This may be because the largest colobus groups, containing many immatures, are attacked most frequently. When a group is reduced in size because its immatures have been depleted, it may become less vulnerable to attack—though this may depend on whether the group's home range is in the chimpanzee core hunting area. J group may not be entirely typical of Gombe groups, in that it lives within this core area; its ratio of immatures to adults is lower than the mean for Gombe.

W GROUP HAD 13 INDIVIDUALS, only 4 of whom were immature, when I began observing it in 1991 (Appendix 19). Buddha, the large high-ranking male, was a fearless defender against chimpanzees, and I witnessed only one successful predation in 14 months of observation between 1991 and 1993. When I returned to Gombe in August 1994 after 10 months away, I could not locate W group. After weeks of following the other four groups, I had not yet seen W and began to suspect that the group, which had decreased in size to only 8 animals in the previous field season, might have been completely eradicated.

Then, only a week before I was to leave Gombe again, I encountered a colobus group of 45 individuals within the range of W group (KK6) that I had never seen before. I recalled that Sharon Watt, studying a colobus group in Busindi Valley, had reported seeing a large, previously unknown group while hiking through my study area. She provided me with census data on the group, which I called NK (New Kakombe).

A major surprise came when I scanned the group and saw Buddha and Bones, the two largest and highest-ranking males of W group, traveling with NK. In my three days of observation, Buddha and Bones rarely interacted with the other males and were quite peripheral to the group as it foraged. It appeared that they were attempting to join NK group and had been only somewhat successful. None of the other members of W group were present, and I inferred that that group had either dispersed or been reduced to a few individuals who were now trying to find new groups. NK group was in an area where only W and J groups had been seen before, in the northeastern section of Kakombe Valley and on the ridge between Kakombe and Kasakela valleys.

I left Gombe that year unsure what had happened to W group and wondering whether these two males would be accepted into an unfamiliar group. Male transfer among red colobus has been reported for only one other population, at Tana River (Marsh 1979). In a male-philopatric species such as red colobus, males are born into the group and therefore are related to some degree, making transfer among groups difficult.

The following year I arrived in July eager to find NK group and to see if Buddha and Bones had succeeded in migrating. However, in several weeks of searching, I was unable to find either NK group or any member of the former W group. It appeared that NK group was not using this new section of its home range during the dry season in 1995. Perhaps that had been a temporary range expansion and the group was now foraging elsewhere. I could not recognize any of the individuals in NK group except Buddha and Bones, so unless they were present I would not be able to identify NK confidently. Apparently Buddha and Bones had either failed in their attempt to immigrate into NK and then dispersed elsewhere or died, or the entire group was foraging in a new area. I favored the former explanation, since I saw groups of the approximate size and composition of NK without either of the two W-group males.

It is unclear why males of one colobus group are excluded from another; by adding a few more adult males, the group might be better able to repel chimpanzee attackers—advantageous even if the new males were not kin. As long as the number of new males was small, the ability of the resident males to fertilize most of the females would not be compromised. Thus a limited number of "hired guns" might be enlisted to help in predator defense. Whether or not this is the case will have to await further study.

Because I could not observe all of the hunts on each colobus group, the only way to gauge the impact of chimpanzee predation on each group was by comparing the levels of hunting by the chimpanzees in each group's home range. I could then note changes in colobus group size and composition from year to year (Appendix 20). The percentage of hunts recorded in each valley corresponded to the amount of time the chimpanzees spent there (Goodall 1986). We have seen that Gombe chimpanzees do not form search images when hunting; instead, they encounter colobus groups opportunistically in the course of daily foraging for plant foods.

Figure 8.5 shows the variation in the size and age structure of red colobus groups from one valley to the next within the hunting range of the Kasakela chimpanzees. The size difference between red colobus groups near the core area of Kakombe and the valleys at the periphery of the chimpanzees' hunting range is nearly 50 percent (Stanford 1995a). The mean size of colobus groups decreases near the core area, creating a concentric effect in which the centrally located groups have compositions strikingly different compositions from those in the more peripheral valleys.

This difference in size is explained largely by the percentage of immatures in each group. Only 20 percent of the individuals in study groups in the chimpanzee core area are immature; in the peripheral groups the figure is nearly 40 percent. Here we have strong evidence of the role that chimpanzees play in controlling the size of the red colobus population. It is also the only case known in which predation can be directly linked with its likely results for a wild primate population.

It is not clear whether the group sizes we see among Gombe red colobus represent evolved responses to the threat of predation or the proximate outcome of continual attacks by predators that keep group size relatively small. The potential utility of a theoretical optimal group size in modeling the predator-prey system at Gombe is doubtful. During the hunting binge of 1990, repeated predations on the ridge between Rutanga and Linda valleys probably rendered groups in that area dramatically different in size and age structure from what they had been a few days earlier. In modeling group-size optimality, few researchers have considered the problem introduced by high predation levels.

Intraspecific Prey Choice and Productivity

By harvesting particular ages or sexes of red colobus, chimpanzees could have varying degrees of impact on the population size and structure. For instance, if they killed large numbers of adult males but not

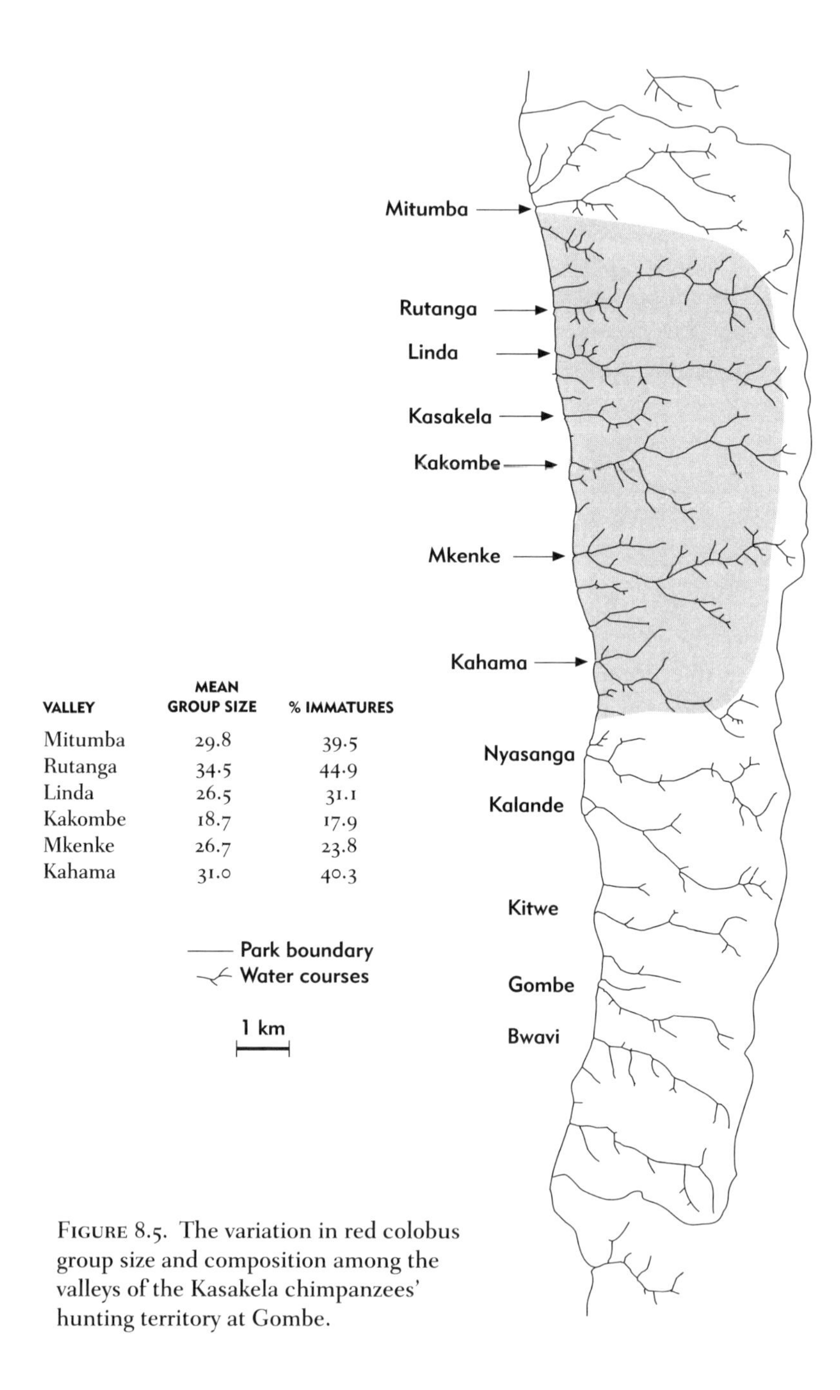

VALLEY	MEAN GROUP SIZE	% IMMATURES
Mitumba	29.8	39.5
Rutanga	34.5	44.9
Linda	26.5	31.1
Kakombe	18.7	17.9
Mkenke	26.7	23.8
Kahama	31.0	40.3

FIGURE 8.5. The variation in red colobus group size and composition among the valleys of the Kasakela chimpanzees' hunting territory at Gombe.

adult females, the demographic outcome for the colobus would be radically different than if females were the main targets, since the loss of prime-age adult females would markedly decrease the productivity of the population.

At Gombe, prey selection by chimpanzees is random with respect to the sex of the victim. The loss of many adults of either sex who have their prime reproductive years still ahead would be damaging to the colobus population. Since many infants die from causes other than predation, their loss can be withstood with less impact on either a female's lifetime reproductive success or on the colobus population as a whole. A very different situation exists in Kibale National Park, where crowned eagles selectively harvest male monkeys and are a possible cause of the strongly female-biased adult sex ratio (Struhsaker and Leakey 1990).

The impact on the Gombe colobus population of killing so many immatures can be seen in what the geneticist Sir Ronald Fisher (1958) called reproductive value. A female's reproductive value is based on the number of potential offspring remaining in her life span—her expected future fitness. This value is low in infant animals owing to their high mortality rate, peaks in the early-prime reproductive stages, and declines afterward until death. The majority of the red colobus killed by chimpanzees are infants and juveniles of low reproductive value. Their lesser contribution to population growth means that a greater harvest of them can be sustained over extended periods of time (MacArthur 1960; Hrdy 1977; Struhsaker and Leakey 1990). If chimpanzees hunted selectively in order to sustain a harvest of prey, they would choose prey ages exactly as they actually do.

Since the interbirth interval among Gombe red colobus is approximately 24 months, each of the twelve adult females in an average-size red colobus group gives birth every 24 months, producing an average of six new infants annually. Documented rates of female immigration add a new breeding female about every three years. If chimpanzees kill about 18 percent of the population overall, 75 percent of which are immatures, about five animals from each group, mainly immatures, are removed from the population each year. Since the infants are replaced as little as 8 months later, it is the loss of the adults that creates problems for the group in sustaining predation. Only one of the two infants remaining from this demographic cohort after predation takes its toll is likely to reach maturity, matching the expected loss of adults shown in the age-specific mortality chart in Table 8.2. However, this estimate is based on the five study groups, which are subject to the most intense predation levels because of their home-range area. For groups living toward the periphery of the Kasakela chimpanzees' hunting range, mortality from predation is much lower and the number of infants surviving each year is higher; the overall impact of predation is therefore substantially less.

The extrapolation above, while speculative, suggests that the effect of predation is not as severe in the long term as might be expected, given such high predation rates. In 1990 and 1992, predation rates were high (with more than 30 percent of the colobus killed in 1992 alone). Other years, such as 1993, were not intense hunting years and colobus mortality was lower. We do not know how often years of intense predation occur, or whether cycles occur that affect the colobus on a regular basis. Rather than ask how the colobus could withstand such intense predation, we may find it more appropriate to ask what would happen to their population size if predation were nonexistent.

Community Ecology

THE ROLE OF THE INDIVIDUAL. Frodo, as we have seen, kills a large proportion (up to 10 percent) of the red colobus population in his hunting range each year. Boesch and Boesch (1989) describe one skilled hunter at Taï whose presence served as a catalyst for hunting by the whole community. One animal of great hunting prowess may affect a prey population profoundly. I call this the Great Chimpanzee model of local ecological history. Because of Frodo's role as a catalyst—many hunts do not begin until Frodo displays an interest—his absence from the Gombe predator-prey system would probably cause an immediate decrease in predation on red colobus.

Prior to Frodo's era as a powerful hunter (before 1987), Goblin was the most accomplished hunter in the Kasakela community (Stanford et al. 1994a). In addition to his own hunting, Goblin used his alpha status to acquire many kills through theft from the captors. In no other year between 1980 and 1986 did hunting approach the frequencies of 1987, 1990, and 1992 (Stanford, unpublished data); the most likely reason for the increase in the late 1980s was the maturation of Frodo.

An intriguing possibility arises. Red colobus affect the forest ecology of Gombe through their harvesting of leaf flush during each growing season. They also contribute to tree germination strategies by dispersing fruits and seeds (as chimpanzees do for fruit trees, Wrangham et al. 1993). Chimpanzees may therefore influence the dynamics of the local community ecology by their annual rate of predation on red colobus. There appears to be a cyclical nature to Gombe predation pressure. It may be produced by the individuals who live in the home range of the colobus during a given decade, a factor that may in turn shape the age structure and size of the colobus population and of each colobus group.

CHIMPANZEES AS KEYSTONE PREDATORS. A number of factors other than predation (such as food availability and disease) also limit red colobus populations. Whether the colobus at Gombe are affected by any food-related ecological density limits is unknown; certainly red colobus elsewhere exist at dramatically higher densities (300 per square kilometer in Kibale, Struhsaker 1975). Chimpanzee predation may keep red colobus density below the level at which food competition would become important in affecting mortality, group size, and population growth. Chimpanzees are therefore the top predators in the Gombe forest ecosystem, even though animal prey are a relatively unimportant part of their diet.

What is important is the predators' impact on the prey species. If chimpanzees regulate monkey populations, then their presence in African forests may be essential to protect the diverse communities of other primates with which red colobus may be ecological competitors, since they too may be controlled by chimpanzees.

Community ecologists debate the most important factors shaping the organization of tropical forests—the top-down versus bottom-up models of community ecology. Bottom-up advocates consider availability of food resources to be paramount in structuring a tropical forest; foliage feeds both vertebrates and invertebrates, which in turn feed larger fruit eaters (Wilson 1987). Top-down advocates see the large carnivores as the key to controlling the populations of midsized mammals, which in turn harvest the vegetation of the forest and shape its structure (Terborgh 1988; Wright et al. 1994).

The keystone-species concept holds that particular species play a controlling role in the ecosystem's structure and function. For example, a particular forest might contain two species of deer that feed on the foliage of different plant species. A large predator, the mountain lion, is also present and can play a key role in determining the ecology of the forest (in turn providing niches for other species) through predation on the two deer species. One deer has twice the reproductive rate of the other and is also more destructive to the forest because of its tendency to browse foliage needed by a variety of other animals. If not for the predator, the more fecund and more voracious deer species would outcompete the other species, perhaps driving it to local extinction. The structure of the forest would be altered, to the detriment of many species in it. But the mountain lion prefers to feed on the fecund and voracious deer, and in doing so maintains a balance between the two deer species. Were the mountain lion removed a sequence of ecological changes would be set in motion that might dramatically alter the network of relationships among the forest's inhabitants.

Are chimpanzees keystone predators? To examine the effect of chimpanzees on the structure of their prey communities, we can compare prey population densities in forests where chimpanzees are absent and in forests where they are present. We can then compare primate communities of which colobus are a part, to see if the absence of chimpanzees appears to change the competitive balance among the monkey species in a way that is reflected in their relative densities.

At least two confounding factors interfere, however. First, like all other predators, chimpanzees vary in their prey preferences from forest to forest, based on both availability of other foods and cultural differences among chimpanzee populations (McGrew 1992; Stanford 1996). Second, some differences between sites that I attribute to the top-down influences of predators might also be attributed to bottom-up influences such as food availability. In this analysis I assume that since Gombe chimpanzees preferentially kill immature colobus, the age ratios of red colobus groups in other sites may be used as an indirect measure of the intensity of chimpanzee predation.

In Table 8.4 I compare the densities and other population parameters of red colobus at eight study sites in East, Central, and West Africa. Four have chimpanzees who hunt red colobus: Gombe, Taï, Kibale, and Mahale. Chimpanzees do not occur at the other four sites: Abuko, Tana, Jozani, and Fathala. The three lowest adult sex ratios of

TABLE 8.4. *Comparison of red colobus density and demography at sites where chimpanzees are present or absent. Sites with chimpanzees are denoted by an asterisk.*

SITE	ADULT SEX RATIO	DENSITY (SQ. KM)	NO. ADULT MALES	NO. ADULT FEMALES	NO. JUVENILES	NO. INFANTS	GROUP SIZE (MEAN)
Gombe*	0.313	42	6.0	11.2	1.8	3.4	23.0
Taï*	.38	66	4.8	12.5	1.0	7.0	40.1
Kibale*	.43	300	4.0	9.3	4.9	5.0	43.9
Mahale*	—	200	—	—	—	—	30.0
Abuko	.18	124	2.2	12.2	4.0	3.9	26.5
Tana	.16	212	1.5	9.6	3.8	2.3	18.0
Jozani	.17	80	3.0	17.7	?	?	37.9
Fathala	.50	—	6.0	12.0	—	—	29.0

Sources: Gombe—this study; Taï—Struhsaker 1975; Kibale—Struhsaker 1975; Mahale—Ihobe, personal communication; Abuko—Starin 1991; Tana—Marsh 1979; Jozani—Mturi 1991; Fathala—Gatinot 1978.

males to females occur at nonchimpanzee sites (Abuko, Jozani, and Tana), and mean group size is lowest at two of the sites (Abuko and Tana). The red colobus populations with the best-documented predation patterns (Gombe and Taï) have the fewest juveniles per group. This is strong circumstantial evidence of the ecological role of predation in red colobus social structure. Red colobus tend to live in groups with fewer males at sites where predation pressure from chimpanzees is absent. The skewed sex ratio is not explained by the absence of females from these groups, since they are preyed upon even more than males are. The number of males per group is also lower at Abuko, Tana, and Jozani. Ecologically, Tana and Fathala are the two driest sites at which food resources might be expected to limit red colobus group size and density (Starin 1991); this is an alternative explanation for the densities observed.

Table 8.5 shows the primate densities in a number of African forests. Only sites in which red colobus are present are included, to test the hypothesis that the presence of chimpanzees structures the arboreal monkey community by removing red colobus as potential food competitors with the other monkey species. The previous table examined whether colobus populations are directly controlled by chimpanzees.

Table 8.5. *African forests in which both red colobus and chimpanzees are present, with population densities of other primates. (Density given in sq. km.)*

			Kibale			
Species	Gombe	Taï	Kanyawara	Ngogo	Mahale	Tiwai
Pan troglodytes	2.5	2.6[a]	>3.0	>2.0	1.0	—[b]
Colobus badius	42	66	300	175	200	50
C. guereza	—	—	45	4.5	1	—
C. polykomos	—	23.5	—	—	—	55
Procolobus verus	—	21	—	—	—	8
Cercopithecus mitis	8	—	45	6		—
C. ascanius	34	—	158	70		—
C. diana	—	17.5	—	—	—	64
C. campbelli	—	15	—	—	—	43
C. petaurista	—	29.3	—	—	—	64
C. l'hoesti	—	—	5	?	—	—
C. nictitans	—	?	—	—	—	—
Cercocebus atys	—	10	—	—	—	39
C. albigena	—	—	10	19	—	—

Sources: Gombe (Tanzania)—this study; Taï (Ivory Coast), Galat-Luong 1983; Kibale/Kanyawara (Uganda)—Struhsaker and Leakey 1990; Kibale/Ngogo—Ghiglieri 1984, Struhsaker and Leakey 1990, Watts, personal communication; Mahale (Tanzania)—Nishida 1968, Ihobe, personal communication; Tiwai (Sierra Leone)—Oates et al. 1990.

a. Taï density before recent epidemics.
b. Tiwai has a very small chimpanzee population.

This table presents data on the possible indirect influence of chimpanzee predation.

At the sites where the density of chimpanzees is highest, the density of red colobus is lowest (with the exception of Kanyawara). However, the population density of red colobus in those forests is also positively correlated with the overall combined density of the other monkey species ($r^2 = .68$, $p = .06$), suggesting that abundance of food and not presence of chimpanzee predators controls population densities. Because red colobus are mainly folivores, while many of the guenon species are primarily frugivores, the degree to which the density of one influences the density of the other probably is limited. It thus appears that chimpanzees are controlling factors on red colobus populations across Africa wherever the two species are sympatric. This predation does not appear to influence the relationships between red colobus and other types of monkeys in these forests.

Modeling the System

In order to understand the potential long-term effects of chimpanzee predation on the red colobus population, we can model a simplified version of the predator-prey ecology. Stella II© 3.0 models the dynamics of natural systems over extended periods. Red colobus reproductive parameters, combined with information on chimpanzee hunting patterns allow Stella to predict the impact of hunting on the colobus population over the next decade. Stella can examine how, for example, a baby boom of male chimpanzees might affect the red colobus population years from now. If chimpanzee hunting patterns are not influenced by overall colobus population size, then predation should have its strongest impact when the colobus population is small and its weakest effect when it is large—or increasing for reasons unrelated to predation.

Modeling results are presented in Appendixes 21–27. If chimpanzees were to disappear from Gombe and the colobus birthrate were to remain constant, the colobus population would rise over the next decade. In ten years the population would be nearly twice its current estimate of 500 within the range of the chimpanzees (Appendix 22). Much of this increase would be due to the higher percentage of the population that is immature and that has been released from high predation levels (Appendix 23). The percentage of the population that is immature would rise from its current 32 percent to 43 percent and then plateau. Note that I have not included any constant to represent a food-resource carrying capacity in this model; it is entirely possible that the colobus population would face limits on growth from food limitations brought about by a population increase. If a small number of adult and adolescent male chimpanzees still existed, this increase would occur more slowly: 5 male chimpanzees would cause the population to rise slowly from 500 to nearly 1,000 over the same ten-year period (Appendix 24).

A population of 15 male chimpanzees would cause a steady decline in the colobus population. Over a decade, continual hunting by many chimpanzees would produce a 40 percent drop in red colobus (Appendix 25). It is unknown whether such a large cohort of male chimpanzees has ever existed at Gombe; Goodall (1986) speculated that the community fission of the 1970s occurred because the cohort of males, then 18, had grown too large. The Ngogo chimpanzee community of Kibale National Park contains more than 20 adult males, and also a lower population density of red colobus than at nearby Kanyawara (Mitani, personal communication).

If the birthrate in the model is adjusted so that female red colobus at Gombe are slightly less fecund (birthrate 0.54 versus 0.40), even a hunting cohort of 8 male chimpanzees produces a steady decline in the colobus population (Appendix 26). A larger number of chimpanzees combined with a low female colobus birthrate leads to local extinction of red colobus. The equilibrium point occurs when the documented colobus birthrate, combined with infant survivorship and other components of individual colobus fitness of breeding, compensates for chimpanzee predation. This occurs at a male chimpanzee population of no more than 8 adult and adolescent males (Appendix 27). This is in fact the mean number of males in the community over the past 30 years, though the range has been 5 to 8.

Models are only as realistic as the available data. This model suggests that recorded wide fluctuations in the age and sex structure of chimpanzee communities have a strong impact on local red colobus populations.

9. Why do Chimpanzees Hunt?

The difference between the ecology of sympatric red colobus and chimpanzees and that of other predators and prey is readily apparent. With less than 5 percent of their diet composed of mammalian meat, chimpanzees are not obligate carnivores and do not appear to rely on red colobus as a source of essential calories or nutrients. Of the four living species of great apes—chimpanzees, bonobos, gorillas, and orangutans—only chimpanzees hunt and consume meat frequently. Hunting appears to be species typical for wild chimpanzees, in that all populations that have been studied hunt intensively. But if meat is such a valued source of nutrients for chimpanzees, then why is there so much seasonal, individual, and annual variation in hunting frequency?

In this chapter I examine the causes of hunting. There is probably no single reason why chimpanzees hunt; of the dozen or more factors that are influential, several may be operating at the same time. Teasing apart these factors consumed my interest for much of the time I studied hunting at Gombe. The ability to predict when hunts were most likely to occur enabled me to observe more and more hunts as the study progressed.

For purposes of analysis, I have categorized the potentially important influences as ecological, nutritional, and sociopolitical.

Ecological factors include aspects of the forest habitat that may mediate hunting patterns in entirely proximate ways. For instance, if chimpanzees undertook hunts in half of their encounters with colobus groups, the density of colobus groups would greatly influence hunting frequency; we would seek the causes of colobus group density to explain chimpanzee hunting patterns. This model of estimating the rate of random encounters has been used to study primate mixed-species associations (Waser 1982; Holenweg et al. 1996). Other factors that promote encounters, such as the daily distance traveled by chimpanzees or the particular paths taken that might lead them into colobus groups, may also influence hunting. These factors are extrinsic to the relationship between the two species.

Nutritional factors may promote hunting as a means to obtain nutrients, calories, or both—available in the carcass of a colobus. Some nutrients, such as saturated animal fat, may be available only from animal prey. They could be trace elements such as calcium or trace minerals, abundant only in animal carcasses. They could even be nutrients obtainable only from colobine monkeys, such as B-complex vitamins that ruminants (and perhaps colobine monkeys) synthesize in their specialized digestive tracts. Sociopolitical factors would promote hunting not because of what a colobus carcass contains nutritionally, but because of what it might represent socially as a tool for barter, for displaying status, and for enticing allies and potential mates.

These categories need not be mutually exclusive: if possession of a colobus carcass has political value, the value stems from its nutrients, without which it would have no currency in chimpanzee society.

Ecological Factors

To investigate patterns of hunting, I started with the null hypothesis: that the timing and frequency of hunting are random, without significant identifiable influences. This hypothesis was easily rejected by the clumped distribution of hunts during the dry-season months of July through September (Chapter 4); the effect is not due to an observer bias in hours spent in the field. Nevertheless, it was still possible that in a particular month hunts might occur randomly, making it important to look for patterns in the occurrence of opportunities that lead to hunts.

The most proximate condition for a hunt to occur is that the chimpanzees encounter a group of red colobus. Gombe chimpanzees do not appear to adopt a search image to locate prey in the way that Taï chimpanzees have been reported to do (Boesch and Boesch 1989). All hunts therefore arise from encounters while chimpanzee parties forage for plant foods. The rate at which chimpanzee parties at Gombe encounter colobus groups, about once every second day (every 24.1 daylight hours, Stanford et al. 1994a), is far less than the encounter rate at Taï (every 1.5 daylight hours, Boesch and Boesch 1989). Because Gombe chimpanzees undertake hunts in a much higher percentage of encounters (> 70 percent) than Taï chimpanzees (< 10 percent) do, the overall hunting rate is greater at Gombe. Gombe chimpanzees hunt more avidly than any other known chimpanzee population (Stanford et al. 1994).

The hunting rate is subject to the distance and direction of chimpanzee group movements as well as the density of red colobus groups and their ranging patterns. Comparison with Taï is not yet possible, in that there is less detailed information on the red colobus population. Ongoing studies of red colobus behavioral ecology at both Taï and Mahale should soon shed light on this issue. Much is known about the red colobus at Kibale, but systematic study of hunting by the chimpanzees has been feasible only recently.

Gombe chimpanzee parties and colobus groups could not be tracked simultaneously with any reliability, but some estimates of the effect of seasonal ecology on ranging are possible. These suggest that encounter rates are tied, at least indirectly, to hunting patterns. Red colobus groups did not travel farther per day during peak hunting months than they did in nonpeak months (two-tailed *t*-test; $t = 1.53$, $df = 7$, $p = .25$), but chimpanzee parties travel farther per day during periods of availability of desired fruits (Goodall 1986). Some of the most-pursued fruits ripen during peak hunting months, bringing chimpanzees into contact with colobus groups more often than would have been the case had fruits not been sought at that time.

For example, during September the fruits of *Uapaca kirkiana* and *U. nitida,* which grow on the highest ridges in Gombe, are zealously sought. During this time the chimpanzees often travel daily from hilltop to hilltop to get them (contributing to the fatigue of researchers who climb after them!). These excursions take chimpanzee parties over great distances each day, including areas of the upper valleys that have large seldom-encountered colobus groups with many immature members. The colobus groups are hunted avidly at this time. In 1990 and 1992, for example, bumper crops of *U. kirkiana* probably contributed to the intense hunting binges that in 1990 nearly exterminated entire groups of colobus living near the ridge between Linda and Rutanga valleys.

Ficus vallis-choudae is a large streamside fig, the fruits of which are eaten by both chimpanzees and red colobus in the dry season. Colobus eat its ripening fruit and leaves, whereas chimpanzees arrive later in the season to eat the ripe fruits. The presence of both ripe and unripe fruit on the same trees creates a magnet for both species. I saw many hunts that began in this *Ficus* simply because the colobus did not avoid the trees in spite of the presence of chimpanzees. Likewise, in September 1992 and September 1994 large stands of *Syzigium guineense* bore ripe fruit in KK5 and in Upper Mkenke valleys, attracting parties of chimpanzees to eat them nearly every day. In 1994 a colobus group came over the ridge from Mkenke to feed in a stand of *S. guineense* in KK5 and was hunted on three consecutive days.

Figure 9.1 shows that two of the three peak hunting months are also those having the greatest overlap in plant-food diet for both colobus and chimpanzees. Some of these species were eaten seldom by one species but were heavily used by the other; in other cases, the plant food was the most-preferred food for both species at the same time.

Figure 9.2 shows that five of the ten most important colobus foods were also major foods for chimpanzees. Peak feeding months for four of the five (*Pterocarpus angolensis, Ficus vallis-choudae, Pseudospondias microcarpus,* and *Syzigium guineense*) coincide with hunting peaks. It therefore seems likely that local fruiting cycles combined with dietary preferences of both chimpanzees and their prey contribute to hunting seasonality at Gombe.

Other physical features of the Gombe habitat promote encounters. In Chapter 2 I disputed some of the most-mentioned human influences at Gombe while proposing more subtle interventions that affect chimpanzee behavioral ecology. Stands of introduced oil palms date to the era before the 1940s, when Gombe became a game reserve. The influence of the fruits of these palms cannot be overstated (McGrew 1992). They are the single most-eaten food of the chimpanzees on a year-round basis (Wrangham 1975), and their distribution affects chimpanzee ranging patterns. They are found mainly on the valley floors, in or near the sites of former villages. The Kasakela chimpanzees appear to spend increased time on the valley floors in order to forage for palm nuts. Since the riverine habitat preferred by colobus is also in the valley bottoms, opportunities arise for encounters. The impression I have gained from spending time with the more recently habituated chimpanzee community in Mitumba Valley is that these individuals spend

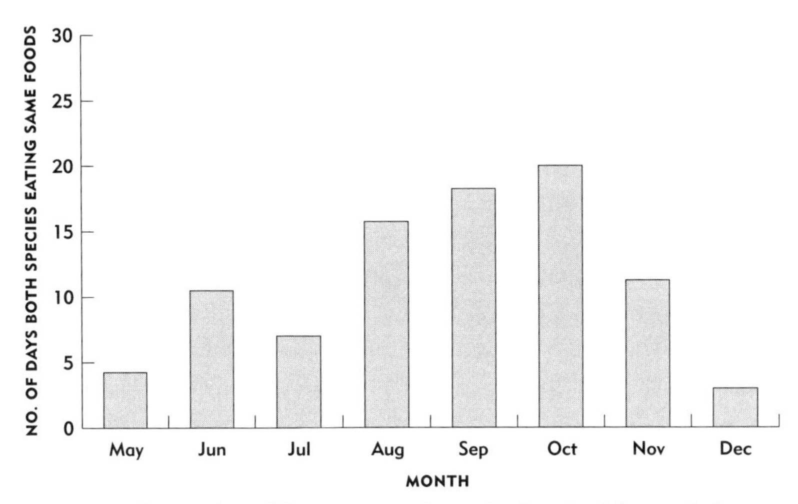

FIGURE 9.1. The number of days per month on which red colobus and chimpanzees were recorded eating the same plant foods at Gombe, 1991–1995.

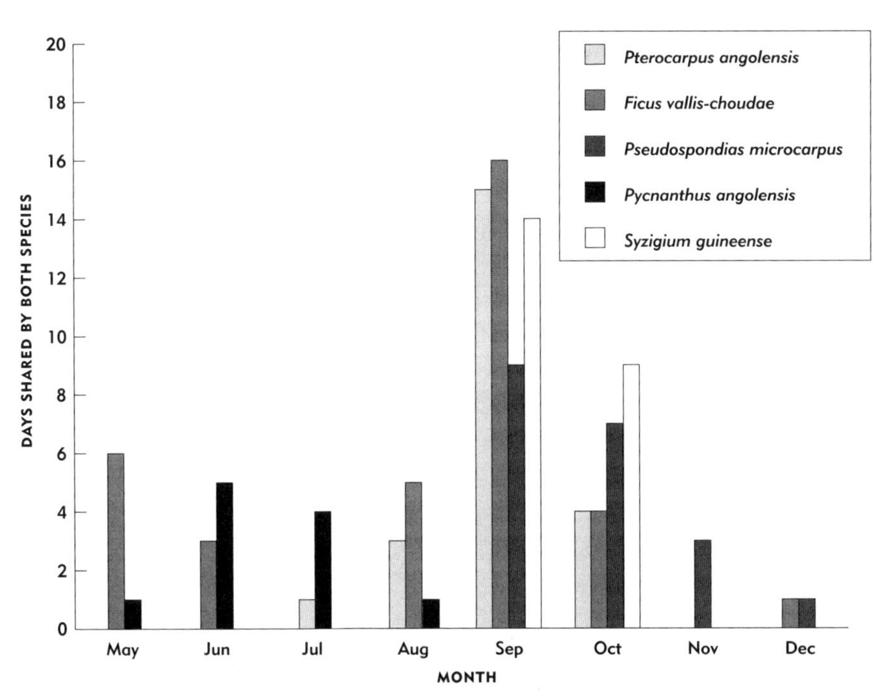

FIGURE 9.2. The number of days per month on which five important shared food species were eaten at Gombe, 1991–1995.

less time in the valley bottoms, which may partially account for their lower frequency of hunting (Gombe Stream Research Center, unpublished data).

Nutritional Factors

Much of the debate about the causes of chimpanzee hunting has centered on the role of nutrition. That chimpanzees eat meat at all is evidence of its nutritional value. Although I have argued (Stanford 1995b) that nutritional explanations are insufficient to explain the observed meat-eating patterns, the reward for the time and energy that chimpanzees invest in hunting is a parcel of fat, protein, amino acids and other nutritious components.

Wrangham (1975) posited nutritional causes in his early study of Gombe chimpanzee ecology, while acknowledging that they seemed to form an incomplete explanation. He reasoned that time and energy invested in the acquisition of meat should be accounted for by the same equation used for other foods: caloric intake must justify caloric output. In the dry season, when Gombe chimpanzee body weights often decrease, meat-eating frequency peaks (though this peak was unknown at the time of Wrangham's study). Thus, an energy deficit may exist and be balanced by the acquisition of calories from animal sources. It has never been shown, however, that nutrient availability is actually lower in the dry season. Even though the forest is nearly bare of foliage during that period, many trees are laden with ripe fruit.

Mahale researchers (Takahata et al. 1984) have long seen nutritional reasons behind chimpanzee hunting. Their rationale is similar with regard to the desirability of animal protein. The Mahale hunting peak comes after the rains have begun in October, but at a time of low availability of ripe fruits; party size decreases in this season. This suggested to Takahata and his colleagues that energy stress was the likeliest cause.

The test of nutritional explanations for hunting rests on the nutrient energy that an individual chimpanzee expects to receive in return for its expenditure of time, energy, and risk of injury. Boesch (1994b) estimated the trade-off in a comparison of Gombe and Taï chimpanzees. He concluded that the average red colobus biomass available to Gombe chimpanzees who hunt alone is high enough that there is no incentive for cooperation among Gombe chimpanzees. The same equation at Taï predicted the high level of cooperation Boesch reported for that population. These estimates were, however, based on comparison of a large sample of Taï hunts against a sample of only 17 hunts at Gombe, all from the months of least hunting activity. The sample is too small to draw any conclusions about the relative hunting incentives for the two populations.

RED COLOBUS VERSUS PALM NUTS AS FOOD. When chimpanzees hunt red colobus, they forgo foraging for plant foods in favor of the quest for meat. Of all the plant foods in the chimpanzee habitat that provide essential nutrients, the fruit of the oil palm is the most valued by both apes and humans. Although the kernel of the nut is highly nutritious, Gombe chimpanzees are unable to crack open its hard shell and so are limited to scraping off the mesocarp. The fibrous mesocarp layer is refined into the palm oil used for cooking in African households, and also in many high-cholesterol baked goods sold in Western markets. McGrew (1992) has described the variation among chimpanzee populations in the use of oil palm nuts. Each tree has fruitless periods (Hartley 1977), yet some trees are always laden with fruit and are relied on by Gombe chimpanzees and other animals. The mesocarp contains more than eight times the calories per kilogram of monkey meat and is higher in saturated fat. The mesocarp provides about 50 calories per nut (extrapolated from Leung 1968). Fat and protein are probably the two components in mammalian meat most prized by both nonhuman and human predators. My work on carcass consumption (Stanford 1996) indicated that Gombe chimpanzees seek fat from the brain and from limb-bone marrow over other body parts after a kill.

In Appendix 28 I present data on the rate at which adult Gombe chimpanzees harvest palm nuts, collected in order to understand the energy trade-offs involved in hunting versus palm-nut foraging. Adult chimpanzees can obtain roughly one palm nut per two minutes of foraging while sitting in the crown of an oil palm. Females and males are not significantly different in the efficiency of their harvesting (two-tailed t-test, $t = 1.64$, $df = 3$, $p = .15$). Individuals varied widely in the speed of extracting nuts, though the amount of fruit remaining in the tree in which they foraged probably influenced the data. (From the ground it is impossible to see the number of palm fruits in the crown).

Even for the slower harvesters, enough palm nuts could be obtained in one hour of foraging (30 palm nuts = approximately 1500 calories) to equal the fat and protein that all hunters except the captor and the most-favored meat recipients would get from an adult colobus carcass. In a study of the nutrients obtained from figs by chimpanzees in the Kibale Forest, Wrangham and his colleagues (1993) found that some fig species provide as many as 40 calories per fruit. It appears that chimpanzees are able to obtain both calories and nutrients from their consumption of plant foods without resorting to hunting for animal prey.

COOPERATION AND HUNTING PAYOFFS. Since a hunt occurs only after individuals decide to take part, we may ask what the expected nutritional payoff will be for each hunter. If the basis for hunting is largely nutritional, then hunters should join parties when the potential payoff warrants the expenditure of time and energy. Success is related to joint participation, so the role of cooperation is a key to understanding why chimpanzees hunt. Cooperative behavior is relatively rare in predatory animals, and apparently evolves only when (1) the net gain to each individual outweighs the costs incurred by having to share the catch with cohunters (Packer and Ruttan 1988) and (2) when the net gain from social hunting exceeds that of hunting alone (Packer et al. 1990).

In early studies of chimpanzee hunting, the small number of hunts observed suggested that group hunting conferred no benefit on the hunters. Busse's (1977, 1978) analysis indicated that success rates actually decreased in larger hunting parties. According to Busse, if joint hunting conferred any benefit, it was that the chaos created by many simultaneous attackers allowed a hunter to capture a confused colobus. Goodall (1986) also considered the confusion effect to be part of the Gombe hunting strategy. No observer who watched chimpanzees before the 1980s saw their hunting as highly coordinated.

Then, in 1989, Boesch and Boesch asserted that the Taï chimpanzees hunted cooperatively and even adopted roles such as "blocker" and "driver" in order to make a kill. Christophe Boesch (1994b) later asserted that the Taï chimpanzees' impetus for hunting cooperatively is the expectation of a reward of meat after the kill. He also argued that Taï chimpanzees hunt more cooperatively than Gombe chimpanzees (Boesch 1994a). These findings on a rain-forest population of chimpanzees suggested that behavioral patterns such as cooperation, long considered crucial to the origins of human cognition, arose in the rain forests of West and Central Africa rather than in the savanna-woodlands of the East. According to Boesch, West African chimpanzees represent the pinnacle, as it were, of chimpanzee behavioral evolution.

The benefits of cooperation in hunting can be profound. George Schaller (1972) found that Serengeti lions doubled their success rate by hunting together. Schaller stressed the difficulty of distinguishing cooperative hunting from communal hunting, in which participants pursued selfish goals. Cheetahs, for example, appear to pursue individual goals during hunts, and their hunting group size is not highly correlated with success rate (Caro 1994). These selfish goals include the ability to kill more prey (Curio 1978), to kill larger prey (Kruuk 1972), to kill more efficiently (Schaller 1972), and to kill with a lower risk of injury to each hunter (Kruuk 1972).

Determining that a behavior is cooperative is not always straightforward. First, many actions performed by two or more animals in concert may appear to be cooperative, when in fact one animal is performing the task and others are simply following. The followers are cheating in that they may reap the benefits of the behavior without active participation. For instance, when lions undertake a communal hunt for a zebra, one lion may take a lead role while others follow but do not actively participate. The followers may obtain zebra meat without expending the same amount of energy or assuming the same risk of injury that the lead hunter does. Distinguishing between cooperation and subtle cheating is not easily done under field conditions. Boesch's descriptions of highly coordinated action and role-taking by Taï chimpanzees appear to depict cooperative behavior. His primary statistical criterion is a widely accepted one: when the hunting success of a group increases as the number of hunters increases, cooperation is implied. This benchmark has been used in studies of cooperative hunting by lions (Packer and Ruttan 1988), African hunting dogs (Fanshawe and Fitzgibbon 1993), and Harris' hawks (Bednarz 1988).

Among Gombe chimpanzees, we know that hunting success increases with party size. Success rates also increase when many adult and adolescent males hunt together (Figure 9.3). This pattern is very similar at Taï, although the relationship is somewhat stronger at Gombe. Paradoxically, no observer who has watched Gombe chimpanzees hunt has obtained empirical evidence or even an impression that they are highly cooperative. Instead, Gombe hunters seem to be

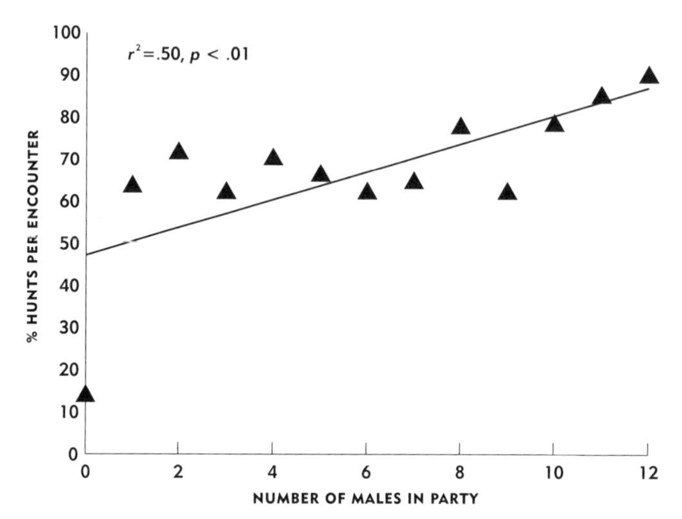

FIGURE 9.3. The relationship between the number of male chimpanzees in the hunting party and the percentage of encounters that led to hunts at Gombe, 1991–1995.

acting individually in a group. The odds of success increase with more hunters simply because the odds are greater that someone will make a kill, not because the combined action of two or more hunters enables a kill to be made. Taï and Gombe are thus similar in the relationship between the number of hunters and their success rates, but the levels of cooperation seem to be radically different (Figure 9.4).

Taï chimpanzees gain at least one much stronger benefit from group hunting than Gombe chimpanzees do. In larger hunting parties at Taï, more colobus meat is caught by the hunting party (Figure 9.5; see also Appendix 29). (This was true at all party sizes in both locations—except when there were five hunters at Gombe, because two of the largest multiple kills were made by this size party.) When a chimpanzee decides to join a hunting party at Taï, he may have learned that more hunters working together means that more kilograms of meat will be available after a successful hunt. The total biomass of colobus killed per hunt is likely to be greater at Taï because of the propensity to capture adult colobus. In a typical kill of three colobus at Gombe, all will be immatures, and the total weight of the carcasses will be 2 or 3 kilograms. A kill of three Taï colobus yields up to three times as much meat because of the larger body size of adults, although multiple kills are rare at Taï compared to Gombe. Boesch (1994b) does not report hunt-

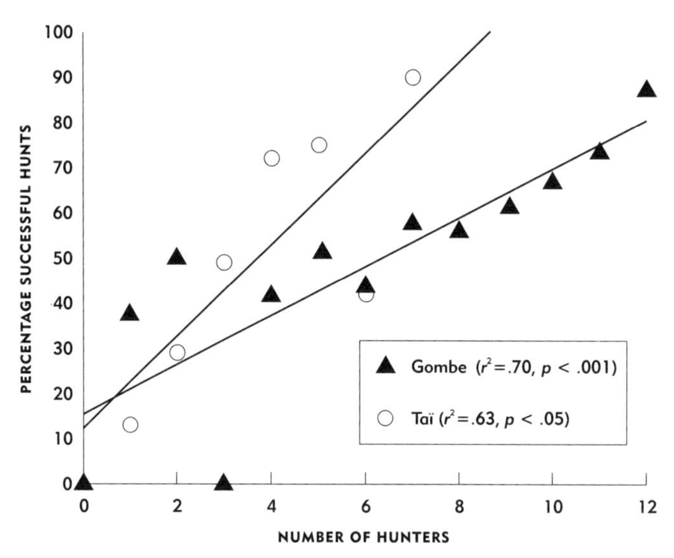

FIGURE 9.4. Comparison of Gombe and Taï in the relationship between the number of chimpanzees in the hunting party and their rate of success. Taï data from Boesch 1994b.

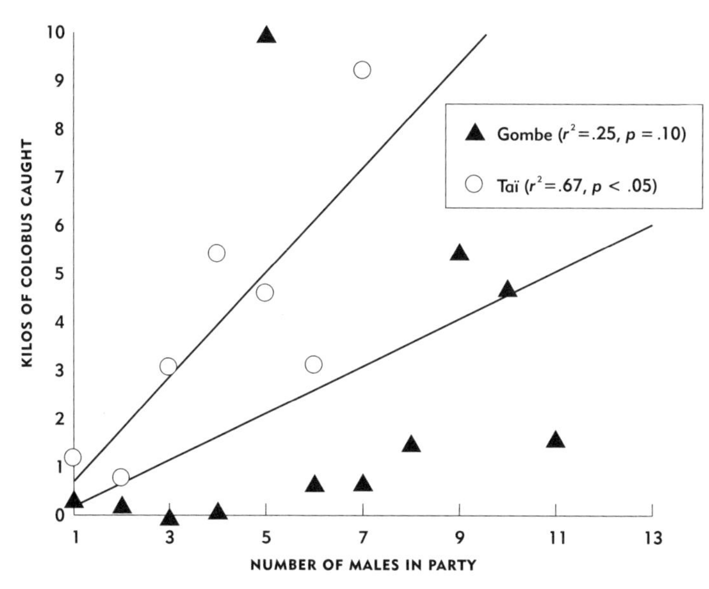

FIGURE 9.5. Comparison of Gombe and Taï in the relationship between the number of chimpanzees in the hunting party and the kilograms of colobus meat caught. Taï data from Boesch 1994b.

ing parties of more than six individuals. The decision to join a hunting party probably is also based on the likelihood of obtaining enough meat for oneself rather than on the likely amount of meat caught overall. A hunter ought to expect to offset whatever time and energy he has invested in hunting instead of in foraging for fruit or other plant foods.

Neither Taï chimpanzees nor Gombe chimpanzees seem to benefit greatly from increased numbers of hunters. Contrary to estimates by Boesch (1994b), the correlation between number of hunters and average amount of meat available to each hunter is not significant at either Gombe or Taï (Figure 9.6). There is, however, a notable difference between the two sites. At Taï, a larger hunting party increases the amount of colobus meat available to each hunter until a fifth hunter joins the party, after which the per-hunter meat total begins to decrease. At Gombe, the return of colobus meat increases when more than seven males hunt together. In other words, Taï chimpanzees seem to benefit from cooperative hunting when in small parties, whereas Gombe chimpanzees benefit more when in larger parties. Note that Boesch defines the hunting party as containing only the hunters, whereas I use the criterion of how many males are present, regardless of whether they all hunt simultaneously.

Boesch (1994c) has claimed that Taï chimpanzees cooperate because the forest canopy is so high that they must do so in order to

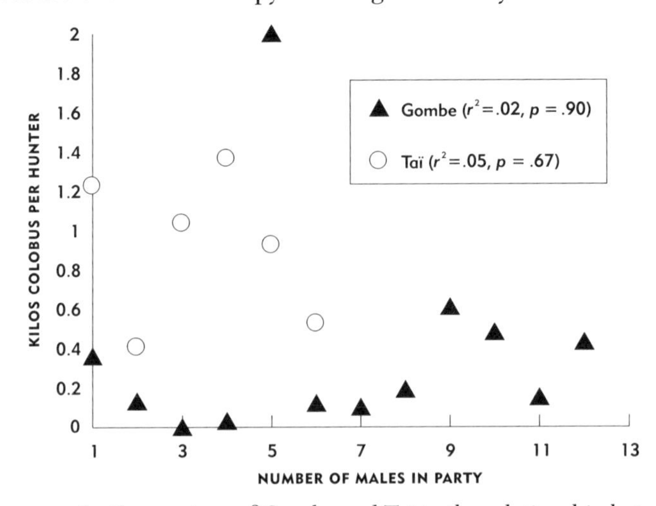

FIGURE 9.6. Comparison of Gombe and Taï in the relationship between the number of chimpanzees in the hunting party and the kilograms of colobus meat available *per hunter*. Taï data from Boesch 1994b.

succeed. The comparison presented here suggests the opposite: larger numbers of Taï hunters would have less incentive to cooperate, for each hunter would have a lower chance of obtaining meat. However, Boesch also calculates that Taï hunters are rewarded for their efforts by receiving a share of meat for their participation, whether or not they have made the kill (Boesch 1994b). Such a system of rewards alters the equation of when to join a hunting party and when not to join. Division of the kill at Gombe follows lines based on alliances, kinship, and dominance (Goodall 1986). The expectation of a meat payoff may not be as critical to Taï hunters if they can expect a share of meat regardless of the biomass of colobus caught. Boesch (1994b) estimates that the nutritional profit margin for the evolution of such hunting is that each hunter must have an opportunity to capture prey weighing at least 1.5 kilograms. At Taï, the modal prey weight is 5 kilograms and the mean weight is 8.7 kilograms (Stanford 1996); at Gombe, the modal prey weight is only 1 kilogram and the mean only 4.4 kilograms.

A primarily nutritional rationale for hunting thus seems likely for Taï chimpanzees, whereas the Gombe data support a stronger social-political basis. I suggest that meat is a currency with different meanings in the two societies. Gombe chimpanzees have taken meat-eating into the larger realm of sex, politics, and manipulation, while this has not occurred at Taï. It must be pointed out again that the value of meat is nutritional at Gombe as well at Taï. Colobus carcasses can be useful as social currency at Gombe only because of their nutritional value. But the chimpanzees there taken the use of meat further; it is a means to an end, rather than only the end itself.

Implicit in the assertions by Boesch and others is the assumption that cooperation in hunting indicates something important evolutionarily because it is a complex behavioral pattern. Cooperation requires the ability to identify and remember partners, to distinguish cheaters from helpers, and to coordinate one's actions with those of others. None of these traits, however, necessarily ranks above selfish manipulation as a hallmark of human intelligence. The use of meat as a tool of nepotism and social maneuvering is as impressive a skill in Gombe chimpanzees as is cooperation among Taï chimpanzees. What is remarkable is the extent to which a scrap of meat that has the same nutritional value everywhere can be accorded different meanings in various chimpanzee societies.

Social Factors

I have argued that if Gombe chimpanzees hunt purely for nutritional gain, they are investing far more time and energy than their nutrient payoff is likely to justify. Now I want to consider the social, political, and sexual explanations for chimpanzee hunting patterns. In order for any of these to be plausible, the protagonist must be a highly intelligent social animal who has the ability to capture other animals, and for whom the meat, blood, and bones of the prey are highly prized.

In the Taï Forest, bushpiglets (*Potamochoerus porcus*) are ignored rather than hunted even though they represent a readily available package of protein and fat. In Gombe, such piglets are gobbled up when stumbled across in their leaf litter nests. Taï chimpanzees eat many adult male red colobus, while at Gombe adult males constitute only 6 percent of the total harvest.

These differences among chimpanzee societies in how meat is used are apparent evidence of cultural variation. The nutritional value of a pig is unlikely to vary markedly from site to site. It is even less likely that a genetic basis for eating pigs exists in one population but not in another. Ecological factors could be involved if the availability of pigs stems from their behavior; perhaps Taï pigs are more aggressive in defending their piglets. The cooperative hunting of the Taï chimpanzees, promoted by the tall forest canopy, may have enabled hunters there to capture adult colobus while Gombe and Mahale chimpanzees are left to pursue immatures. But the possibility remains strong that such differences exist because the current generation of hunters follow hunting traditions learned through years of observation of other hunters.

Social and political explanations for chimpanzee hunting are not new. The earliest modern hypotheses to explain hunting stressed social display by males as the rationale for catching other mammals (Kortlandt 1972). Fifty years ago, Robert Yerkes (1941) considered the chimpanzee male-female relationship to be akin to prostitution, in that captive females rendered sexual services to males in exchange for shared food. Teleki (1973) invoked social networks and inferred some reproductive benefits from his observations that male chimpanzees at Gombe shared meat preferentially with females who had sexual swellings. Nishida (1996) has reported similar behavior by the long-reigning alpha male Ntologi at Mahale. Nishida and his colleagues

(1992) showed that Ntologi used meat as a political tool to solidify his alliances and to snub his rivals. His meat-sharing increased just after his return to alpha status following an overthrow, and declined when his hold on the alpha status was once again secure. Surely a social strategy in his pattern of meat-sharing may be inferred. McGrew (1992) showed that Gombe females who received the largest shares of meat after kills also had the highest number of surviving offspring.

Although hunting at Gombe is seasonal, hunting binges can occur in any month and may last for days or weeks. They tend to happen most often during August and September, but they take place at other times of year as well. Janette Wallis (1997) pointed out that female chimpanzee reproductive cycles show seasonal sexual swellings even though births are not seasonal. Females swell not only when they are ovulating but also when pregnant and lactating, though these anovulatory swellings are less regular in occurrence. The average number of females with sexual swellings peaks in the same months that hunting also peaks (Stanford et al. 1994b), and swollen females are a driving force in the social system. When females are in estrus they attract males, creating large parties that last for days until the males lose interest in the detumescing females.

Our analysis of the relationship among swollen females, party sizes, and hunts over a 10-year period reveals a strong positive correlation among all three (Stanford et al. 1994b). The best predictor of hunting is the presence of at least one swollen female in the party. This is true after other major factors, such as the number of males present or the size of the party overall, are held constant. When a foraging party of chimpanzees passes under a tree that holds a group of red colobus, it seems that males make a series of individual decisions to hunt based on the likelihood of success and of expected nutritional and social gains: How many young colobus are there? How many hunters are available to catch them, and what is my likely share? Is Frodo or another highly skilled hunter present? Are there particular females about who will respond to the lure of meat? Are there males present whose favor needs to be curried with a share of meat? Such decisions may involve high-level associative reasoning, or they may be no more sophisticated than the daily decisions of many other less intelligent social mammals. But sociosexual factors are strongly implicated in the explanation for observed hunting patterns, at least at Gombe.

If the presence of swollen females is a strong influence on hunting by males, then it may be that males use meat from the colobus to obtain more matings from females. The reproductive strategy is not entirely effective, since female chimpanzees are at times highly promiscuous. The use of meat as an enticement for females to barter sex may be a tactic used by males to gain additional or timely matings from preferred females.

Table 9.1 presents observations of instances in which adult males traded colobus meat for matings. In 33 percent of all successful hunts in which both adult males and swollen females were present, meat was exchanged along with the matings. In some instances it was not obvious that colobus meat was being exchanged directly for mating, because often a lot of both was occurring at the same time. But in five cases the sequence was unambiguous; the photo insert shows a meat-for-mating bout. In all but one of these episodes the female was fully swollen. Because female chimpanzees in the wild very rarely mate when not fully swollen, Sandi's 1992 copulation with Goblin was unusual.

Three of the five cases followed the kill of seven colobus on 7 October 1992 in KK6. Atlas captured a juvenile colobus who leaped into his hands while trying to escape. He killed it with a bite to the neck, and within seconds Trezia, a recent female immigrant into the Kasakela community, ran to him and begged for meat. Atlas held the colobus carcass behind his body out of Trezia's reach, whereupon she turned and presented her swelling. They copulated, then parted; Trezia turned and extended her hand again in the chimpanzee begging gesture. Atlas waved the colobus carcass in front of her, then placed it behind his back again until she turned and they copulated once more. After this

Table 9.1. *Observations of matings at Gombe that were exchanged directly for colobus meat.*

DATE	MALE PARTICIPANT	FEMALE PARTICIPANT	FEMALE SEXUAL STATE
7 October 1992	Atlas	Trezia	Fully swollen
7 October 1992	Beethoven	Gremlin	Fully swollen
7 October 1992	Goblin	Sandi	Not swollen
16 October 1992	Wilkie	Gremlin	Fully swollen
12 November 1992	Beethoven	Jiffy	Fully swollen

mating he gave meat to her and also to his younger brother, Apollo. Moments later, Wilkie discovered the pair sharing the carcass and routed them, chasing Atlas and viciously grabbing the carcass when Atlas failed to surrender it immediately. Wilkie then carried the stolen carcass to the limb on which Atlas had just sat. Trezia now directed her begging to Wilkie, who shared the carcass with her. Almost immediately Beethoven, who had captured his own colobus only a few meters away, engaged in the same mating and sharing sequence with the swollen Gremlin. Later in the four-hour meat-sharing episode, Goblin used as part of his courtship display the carcass of an adult colobus he had stolen from the female Sandi. A male chimpanzee often signals his interest in copulating by shaking a nearby branch or bush or by slapping his thigh within sight of a swollen female. (Wilkie has often used the latter signal.) In a rather graphic illustration of the connection between meat and mating intentions, Goblin slapped the inside of his thigh with the limbless and headless carcass of a female colobus; Sandi responded by allowing him to mount her and subsequently received colobus morsels.

Putting the Explanations Together

The sociosexual basis for hunting is a part of the political arena that includes dominance. Dominance status influences which male chimpanzee will obtain the most matings from swollen females (Goodall 1986) and also which male will be able to steal meat with impunity, potentially drawing even more females to him. These factors derive from the value of meat as a nutritional supplement in the diet. Females desire meat as much as males do, but are constrained from hunting for it themselves. Attacking male colobus is a difficult, dangerous activity for females carrying dependent offspring, a likely reason why females rarely make kills.

Although researchers have tried to disentangle the complicated ecological, nutritional, and social causes of hunting, it may not be logically correct to do so. In Figure 9.7 I show the relationship among the factors most directly influencing the frequency with which male chimpanzees at Gombe hunt.

Other than the availability of other foods, the lower boxes are influenced by the ecology of colobus or chimpanzees or both: shared resource use, encounter rate, day range, group density. The boxes at

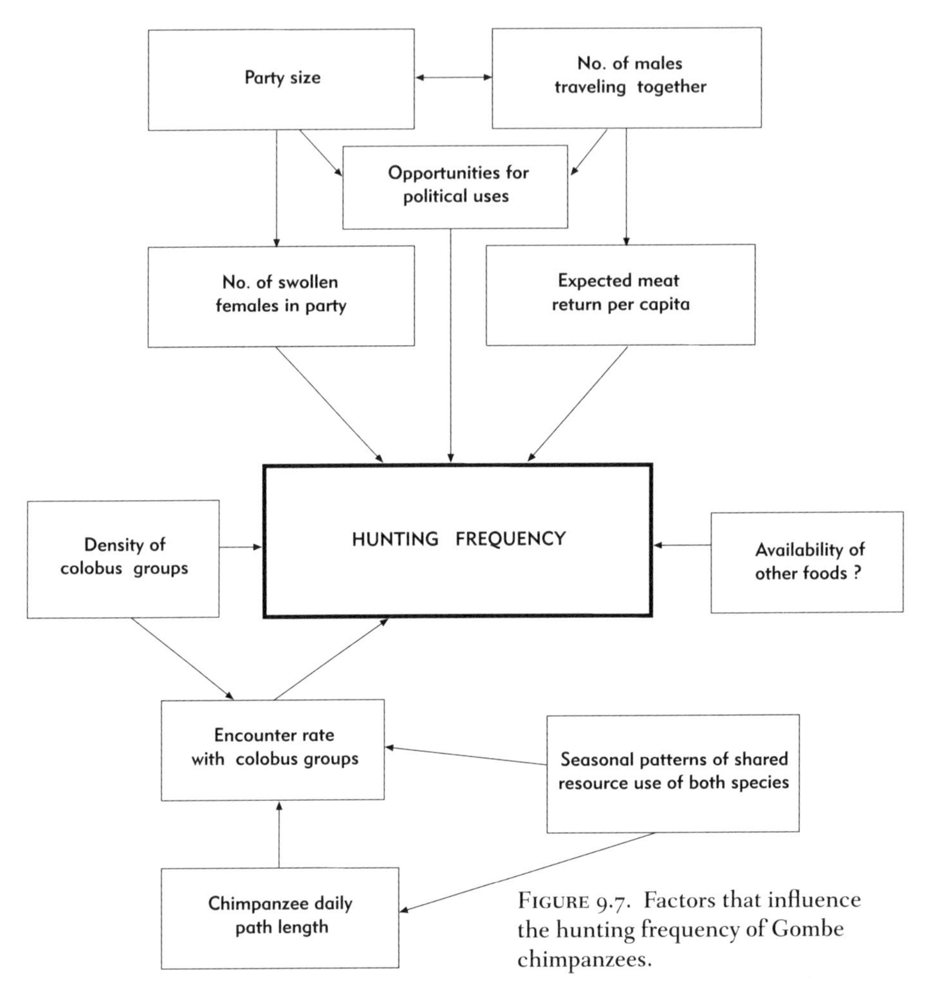

FIGURE 9.7. Factors that influence the hunting frequency of Gombe chimpanzees.

the top are social factors: party size, number of males in the party, estrous females in the party, per-capita return rates, and opportunities for political uses of meat.

Ecological influences change annually depending on the fruit crop, and seasonally depending on the distribution of ephemeral food sources. Social factors also change annually owing to long-term demographic patterns and to yearly differences in the number of females cycling. The two primary influences on party size and number of males per party are the number of estrous females and the availability of fruit patches; these two categories are interwoven and have not yet been separated for any chimpanzee study site. At Gombe, years in which many females are cycling are also years in which party sizes are large (Goodall 1986) and in which hunting is frequent (Stanford et al. 1994b). The 10-week hunting binge in 1990 occurred when seven females were swollen, more than had been recorded at Gombe in the previous two decades.

Scavenging and Its Absence

Since meat is strongly desired as a nutritional supplement in the chimpanzee diet, why do chimpanzees not scavenge more frequently? Scavenging has been reported at Gombe (Muller et al. 1995) and Mahale (Hasegawa et al. 1983), yet it occurs very rarely among wild chimpanzees. Most instances of eating meat that the chimpanzees have not themselves killed occur when fresh carcasses of prey are pirated from baboons (Morris and Goodall 1977). In 36 years of research, the total number of observations of passive scavenging of carcasses encountered is less than 20. It should be noted that scavenging evidence can only be distinguished from hunting evidence by direct observation, a condition that requires habituated chimpanzees. Thus, studies of unhabituated chimpanzees that rely on dung analysis may include unknown levels of scavenged meat as well as live prey.

McGrew (1992) argued that chimpanzees would scavenge more if they lived in habitats conducive to finding uneaten carcasses, such as savannas rich in both ungulates and large predators. But when chimpanzees encounter the fresh carcass of such an animal, they usually regard it with only mild curiosity and consume only a small amount (Goodall 1986; Stanford, personal observation). It is hard to reconcile the obvious desire for meat, bone, and any other part of a killed carcass with the ambivalence to a recently dead but equally nutritious animal.

Scavenging may be a rare occurrence because the chimpanzees avoid the pathogens present in a rotting carcass. Most traditional human foragers cook their meat, but human foraging societies sometimes also eat uncooked meat from long-dead animals (Blurton-Jones, personal communication). Cultural factors too may be at work. Just as Taï chimpanzees ignore pigs, perhaps Gombe chimpanzees ignore most scavenging opportunities because they have little experience eating carcasses. Gombe in particular is a habitat lacking in large mammals, though several ungulate species occurred there in the recent past. If the older males in a hunting party show disdain for the fresh, perfectly edible carcass of a bushbuck, then the younger hunters will most likely do the same. In this way the absence of large ungulates dying in the Gombe ecosystem may indirectly affect scavenging, in that a culture of scavenging would be unlikely to develop.

Why Don't Bonobos Hunt More Frequently?

Related to the role of traditions in chimpanzee hunting is the question why bonobos (*Pan paniscus*) hunt so rarely compared to chimpanzees. The two are close relatives, having diverged from a common ancestor only some 2.5 million years ago (Ruvolo et al. 1991). There are only a handful of records of meat-eating by wild bonobos (Kano 1992; Hohmann and Fruth 1993). Since colobine monkeys occur at the two longest-running bonobo study sites, it is difficult to understand why bonobos show little interest in hunting in light of what a prized nutritional resource meat is to chimpanzees. Reports exist of bonobos grooming colobus monkeys (Ihobe 1992) and catching guenons (*Cercopithecus ascanius*) and black-and-white colobus (*Colobus angolensis*)—but using them as playthings instead of food (Sabater Pi et al. 1993).

Party sizes in bonobo populations tend to be larger than those in chimpanzee populations (Chapman et al. 1994), which should enhance hunting success if bonobos were to hunt in the same manner as chimpanzees. Two decades of research at three sites have produced only eight direct observations of predation by bonobos (Uehara 1997). The lack of complete habituation at Lomako and the use of provision-aided observations in an artificial forest clearing at Wamba may have resulted in some bias against observations of hunting, but it is evident that bonobos do not hunt with the same enthusiasm as chimpanzees.

The lack of a desire for meat seems nutritionally inexplicable. Unless one posits that bonobos have very differential dietary constraints, they should avail themselves of meat-eating opportunities as often as chimpanzees do. Wrangham and Peterson (1996) have suggested that bonobos are generally less aggressive than chimpanzees, a fact that may explain their failure to hunt (although these researchers also point out that hunting and aggression are not necessarily linked).

The data supporting lower levels of male aggression among bonobos are, however, equivocal; bonobos engage in aggressive (but not lethal) intercommunity territorial encounters. It is unclear, in any case, whether predatory behavior constitutes aggression; many unaggressive social animals are efficient hunters. I argue that many of the differences attributed to these two species are the result of insufficient field data on bonobos combined with an eagerness to view that species as a divergent path to modern human behavior (Stanford, in press).

The lack of hunting by bonobos is explicable, however, if hunting has a strong sociopolitical basis. Female bonobos are dominant to males in feeding situations (Kano 1992) because of strategic male deference to females in order to obtain matings (Wood and White 1996). Instead of begging for meat, females simply take it away from males as part of their proclivity to control food resources. Takayoshi Kano (1992) saw females taking carcasses from male captors with impunity, which is a very rare occurrence in chimpanzees. This observation supports a social rather than a nutritional explanation for hunting in chimpanzees. Male chimpanzees may use meat for social manipulation of their groupmates, but this strategy would not be very effective for male bonobos in that they wield less power over females of their species. Males would be unlikely to hunt for prey if the prey were often simply taken from them after capture.

During any given day, an animal may fail to obtain a meal and go hungry, or it may fail to obtain matings and thus realize no reproductive success, but in the long term, the day's shortcomings may have minimal influence on lifetime fitness. Few failures, however, are as unforgiving as the failure to avoid a predator: being killed greatly decreases future fitness.

—Lima and Dill (1990)

10. Predation and Primate Social Systems

Every wild primate dies after living a life of near-constant peril. Natural selection filters the behavioral options the animal can choose at any moment of that lifetime, rewarding good choices and punishing bad ones. How can natural selection choose one specific physical trait or behavioral tendency over all others in such a complex system? The assumption that behavioral ecologists often make is that all traits are adaptive, in that their consequences influence reproductive success. The reason natural selection can mold behaviors that are crucial to survival is the scale of time: the filtering of good and bad traits happens slowly enough that it is beyond the scale of human experience and observation. But it occurs nonetheless.

The failure to gather detailed information on the causes and consequences of predation in wild populations is one of the most notable gaps in the field of primate ecology. The nature of predation is such that it occurs infrequently, at least from the perspective of the human researcher, and what we know about its workings comes largely from circumstantial evidence and compilations of anecdotes. Its importance is often underestimated until the appearance of evidence that cannot be ignored. For example, Sarah Hrdy's (1977) reports of Hanuman langur infanticide quickly led to other studies that documented infanticide in a variety of species. Only in the past decade have hypotheses been tested under field conditions, usually employing taped playbacks of predator calls to observe the prey's behavioral response (Hauser and Wrangham 1990; Noë and Bshary 1997).

In order to comprehend the role of predation in shaping primate social systems, we need both more field data and conceptual models to interpret our data. The presence of predators must be evaluated for its effect on group size, age and sex composition of the group, and ranging, feeding, and grouping patterns of the prey.

In this chapter I offer a conceptual framework for interpreting the effects of predators on the evolution of primate social systems, in particular on the benefits of male sociality in primate groups. Using a comparative approach, I examine the functional similarities between male sociality in intergroup encounters and in antipredator protection. These are the grave risks faced daily by wild primates and other social mammals. Most conceptual models that compare the significance of feeding competition with predator avoidance consider these risks to be equivalent. I propose that the grave risks involved in predation have an effect on fitness fundamentally different from the risks associated with foraging. First, I argue that the influence of grave risks is underestimated and that of chronic risks overestimated. Second, I hypothesize that male sociality evolved in many cases for protection against the risks of both intergroup aggression and predatory attacks.

How Predation Affects Reproductive Success

Primate reproductive success is affected by mating access, by feeding exigencies, and by avoidance of predation. As Lima and Dill (1990) note, failures in the first two of these are more forgiving than in the last. In most species of higher mammals, innumerable copulations occur for each conception; any single mating opportunity therefore is noncrucial. Likewise, feeding opportunities lost tend to be chronic in occurrence and minor in long-term effect on survival. But predation is unforgiving; one failure is the end of all future reproductive chances. This statement of course applies only to the successful predatory episodes, which are a minority of total attempts. Aside from experiencing death at the hands or teeth of a predator, the prey animal may face other disturbances to its life and reproductive goals from predation pressure. These disruptions may carry minor costs that necessitate expending small amounts of energy to compensate by foraging further.

We know very little about the actual energy balance of wild primates, but it is likely that daily foraging occurs within fairly tight energy constraints (Altmann 1980). The intensity of these constraints varies seasonally and annually. The chronic problem of food competition may grow acute when the local food supply crashes (Milton 1982), but aside from such catastrophes, food procurement is a matter of continual low-level risks and gains. The bonds that exist between related females of a primate group have been linked to food defense (Wrangham 1980), whereby females in groups that are able to dominate resources increase their individual reproductive success. This model has stimulated much further research on male and female bonds, although its main points have not been well supported by field evidence. Such intergroup food competition is probably best considered in the category of episodic low-level risks; what is at stake is typically access to food that can be dominated by one group at the expense of another (Figure 10.1). Other costs of fitness are more damaging, though still rarely lethal: fights that carry a risk of injury, recurring nonlethal diseases, and parasite load. Still other risks exist that are potentially lethal: encounters with marauding conspecifics, attacks of disease epidemics, and the risk of infanticide.

An attack by a predator may happen at any moment. It is a catastrophic loss of all future fitness and need occur only once in a lifetime. The threat of instant death and loss of fitness is a grave risk, the rarity of which belies its disastrous outcome: the forfeiture of all future reproduction. In addition, although predation presents an irreversible loss of reproduction to the victim itself, the kin of the victim also suffer a loss of lifetime inclusive fitness. The possibility of death to the animal

	CHRONIC RISKS	EPISODIC RISKS
LOW-LEVEL RISKS	Intragroup food competition, parasites	Intergroup food competition
GRAVE RISKS	Chronic diseases	Predation, attacks by conspecifics, infanticide, food resource crashes

FIGURE 10.1. Mortality risks facing wild primates and other social mammals.

at all stages of the reproductive life span, added to the summed risk of death to all close kin, surely produces a strong selection pressure for predator avoidance. Because predation represents a cessation rather than an incremental loss of fitness, natural selection must exert a stronger force for predation avoidance behaviors than for increased food procurement. The infrequency of predation may have prevented primate ecologists from distinguishing among these different risk levels.

Moreover, while most wild primates consume a wide variety of food types and food species for which there is no single best foraging strategy, many predatory populations specialize on a single prey species (Caro 1994, Serengeti cheetah and Thompson's gazelle; Mech 1970, Isle Royale wolves and moose). In such cases the predator may have evolved behavioral adaptations designed specifically for one prey species, thereby increasing the risk of capture and increasing selection for the prey's detection and escape adaptations. Episodic risks such as predation would thus be expected to affect the same components of female fitness as more chronic risks.

Other Sources of Mortality

Grave risks other than predation affect fitness. The threat of attack by the members of a neighboring group exists for a number of primate species. Such attacks often occur near territorial borders. Encounters between groups can be over territory, food, mates, or all of the above. Among primates, only chimpanzees and humans engage in lethal intergroup aggression, but lethal encounters occur also among other social mammals such as carnivores. Nonlethal encounters are common among a wide range of nonhuman primates.

Another source of conspecific-induced mortality is infanticide. The degree to which infanticidal behavior is widespread in the primate order has been debated (Bartlett et al. 1993; Hrdy et al. 1995). In some species, such as Hanuman langurs (*Presbytis entellus,* Hrdy 1977; Sommer 1994), red howlers (*Alouatta seniculus,* Crockett 1984), and mountain gorillas (*Gorilla gorilla beringei,* Watts 1989), infanticide is an important cause of mortality to immatures and occurs in a context of sexually selected aggression. Infanticide in Hanuman langurs occurs when extragroup adult males oust the resident male and proceed to kill one or more of the group infants. Hrdy (1977) hypothesized that females are selected to accept the presence of infanticidal males because they possess a behavioral trait that makes them more reproductively fit than noninfanticidal males. In most infanticidal species, the sample of observed deaths is too small to test hypotheses about functional causes. Infanticide has even been invoked as an evolutionary force favoring sociality in at least one primate group—gibbons—in which the behavior has never been reported (van Schaik and Dunbar 1990). Although infant-killing poses a grave risk to immatures in some species, the fitness cost to mothers who lose infants may be low enough for the behavior to have evolved and been perpetuated.

Two final sources of mortality for all wild-animal populations are disease and parasites. While the majority of pathogens probably have a minor impact on lifetime fitness, periodic outbreaks of lethal epidemics can cause high mortality. Both the risk of a lethal parasite load and the risk of disease (Freeland 1976) have been suggested as influences on group size and the degree of sociality in nonhuman primates. The degree to which disease factors are significant determinants of fitness is difficult to evaluate, since wild animals that succumb to disease may die of other causes: they may fall prey more easily or become unable to forage and therefore starve to death.

Predation Risk or Predation Mortality?

Predation risk is the chance that a prey will be captured during an encounter with a predator. It is unlikely that prey species have evolved the ability to evaluate precisely the risk factor they face in each circumstance of their daily life. A number of factors should covary with the degree of risk of being captured by a predator: the openness of the terrain (Crook and Gartlan 1966), the density of predators of various species (Abrams 1994; Stanford 1995), the centrality of the animal in a social group (Collins 1984), its body size and age, and its grouping pattern. Since all of these variables fluctuate, natural selection should favor risk-level estimation that takes into account vulnerability to a predator in relation to the quality of the food patches that are at hand. It may be that a primate can roughly estimate the apparent risks. Alternatively, it may overestimate the risk to compensate for dangerous unknown factors, such as the density of predators in the area it must enter to forage (Abrams 1994).

Researchers have debated whether the main agent of natural selection that acts via predation is the risk of being killed or the mortality itself. Geerat Vermeij (1982) argued that only unsuccessful predatory attempts could lead to the evolution of antipredator traits, since the traits that enabled the escaping prey to avoid capture would then be left in the population. Other members of the population would have their genotypes eliminated. In most predatory mammals, however, hunting success rates are lower than 50 percent, meaning that for most prey unsuccessful predatory attempts are the rule rather than the exception. If there is variation in the ability of prey to escape capture by predators that is genetically based, then beneficial traits ought to be selected. Antipredator strategies should vary from one population to another in relation to the predators present and the local habitat features. For primate populations in which predation exerts a strong pressure, selection operates by removing genes from the population. The survivors confer on their offspring whatever antipredatory traits they have evolved. Therefore, both mortality and risk operate simultaneously.

Most animals born into a wild population will die without reproducing, so mortality culls ineffective counterpredatory behaviors. This process is especially apparent when death occurs before the prey reaches reproductive age. Animals who are not vigilant are more likely to die young than those who scan assiduously for predators and listen carefully for alarm calls (Lima and Dill 1990). Selection's impact is less important when predators kill older prey who have completed or nearly completed their reproductive life span.

Age-specific predation patterns can also influence the behavior of prey animals when attacked. By either protecting or failing to protect group members, adults make behavioral decisions under imminent threat of death according to the reproductive value of the group members most at risk. Adults may not try to mitigate the mortality risk to immatures, who have low reproductive value, if doing so entails grave risk to themselves (Clark 1994).

Perhaps the most common effect of predation risk is the disruption of a wild primate's activity budget. For instance, a primate group avoids a rich food patch because it sees a bird of prey soaring overhead or because it hears alarm calls from other species emanating from a thicket. A distinction between predation risk and predation-induced mortality is that coping with the risk of predation may involve only brief and temporary disruptions of the activity budget. Mortality itself causes the loss of many reproductive opportunities.

Male Sociality in Primate Social Systems

It has become obvious that while most early studies of nonhuman primate societies reported mainly societies in which males dispersed while females remained in their natal groups (female philopatry), many species either are variable in their dispersal pattern or are male philopatric (Moore 1984; Strier 1994). Among the male-philopatric species are some that can also be classified as male-bonded (Table 10.1). The distinction is meaningful: male philopatry denotes a society from which females emigrate but in which males are not necessarily tolerant of one another. In male-bonded primate societies, males are usually (but not always) related to one another and form long-lasting alliances that comprise mutual tolerance, cooperation, and coordinated action. For example, among hamadryas baboons, even though the females transfer between one-male groups, the males of different groups are not bonded to each other (Kummer 1968). The one-male groups are discrete entities but interact in higher-level associations with other one-male groups.

Primate societies can be considered nuclei of females around which males aggregate, and these female cores may become defensible by males (Emlen and Oring 1977; Wrangham 1980). Still, females of some species do not stay in a cohesive and defensible group; they disperse to forage and then may be defensible only through the concerted effort of males (van Hooff and van Schaik 1992). Chimpanzees are an example. Male bonding has often been considered a coalitional strategy of males to control reproductive access to females, among both primates and other large mammals such as lions and cheetahs (Packer et al. 1988; Caro 1994).

Mutual male tolerance may also be related to each male's ability to aid females in controlling patchily distributed food resources, so that groups of females tolerate the presence of multiple males who are providing them with a service (Wrangham 1980). And, as we have seen, male bonding can be effective in antipredator defense, when joint counterattack helps to repel predators and thereby enhances the fitness of the participating males (Busse 1977; Boinski 1994).

If male sociality has benefits that allow males to control females, why is it that the males of more primate societies do not form close bonds? For instance, male bonobos remain in their natal community but do not form strong bonds. Male control of female reproduction is not a feature of bonobo society, as it is among strongly male bonded chimpanzees (Furuichi and Ihobe 1994). Males of other female-philopatric primate species form fleeting bonds to achieve a goal such as mate competition and control, but the bonds dissolve once the goal is achieved (Hanuman langurs; Hrdy 1977; baboons, Noë 1992). In many species characterized by multimale groups, unrelated males cooperate in many of the ways male kin would, such as antipredator behavior (van Schaik and Hörstermann 1994) and the repelling of extragroup males (*Cebus capucinus,* Rose 1994; Perry 1996).

Male bonding is associated with active antipredator defense in about half of the primate populations listed in Table 10.1. All of the species characterized as male bonded (MB) feature cooperative male defense against predators. Note, however, that cooperative antipredator defense by group males also characterizes about two-thirds of the populations not considered to be male bonded. Nearly every species (13 of 15) that lives in groups with male-biased sex ratios engages in cooperative male defense against predators, while 15 of 20 species with sex ratios that are not biased also do so. In other words, male bonds are a typical feature of those species in which males actively repel predators, but even in species without male bonds, active antipredator defense occurs. We can conclude only that male bonds in primate societies are sometimes, but not necessarily, features associated with predation pressure (see also Stanford in press a).

The intensely male cooperative behavior that characterizes a minority of primate species has probably evolved for two reasons: first, male affiliation is an effective response to predators; second, males may be able to defend the group against attacks from neighboring male-bonded groups that attempt to gain access to the group's resident females (Table 10.1). Male bonding is advantageous because these external threats can be countered best by the combined efforts of several individuals. If physical confrontation of predators is unavoidable or is the most effective strategy, then a joint response by multiple males may be the best antipredator defense (Boinski 1994).

It has been theorized that male bonding evolved to enhance the control of females too dispersed or too numerous for males to control indi-

TABLE 10.1. *Male-philopatric and male-bonded primate species.*

SPECIES	MALE: FEMALE RATIO IN GROUP	PHILOPATRY AND BONDING PATTERNS	REPORTED MALE ROLE IN PREDATOR DEFENSE	REFERENCES
Lemur catta	1.13	MT, FB	Cooperative multimale defense	1
Callithrix geoffroyi	1.00	Not known	Cooperative mobbing	2
Callithrix jacchus	1.00	MT, FT		3, 4
Saguinus fuscicollis	1.75	MT, FT	Cooperative mobbing	5, 6
Alouatta palliata	0.38	MT, FT	Not reported	7, 8
Alouatta seniculus	0.60	MP, FT	Multimale defense	9
Ateles paniscus	0.32	MP, MB	Occasional multimale defense	10
Brachyteles arachnoides	0.78	MP, MB	Occasional multimale defense	11
Cebus apella	0.89	MB, MT, FT	Multimale mobbing	12, 13
Cebus capucinus	1.38	FB, MB	Multimale mobbing	14, 15
Saimiri oerstedii	0.63	MP, MB	Multimale mobbing	16, 17
Saimiri sciureus	0.30	FB	Multimale mobbing	18
Saimiri boliviensis	0.30	FB	Multimale mobbing	19
Colobus guereza	0.33	MT, FB	Multimale defense	20

Table 10.1 *(continued)*

SPECIES	MALE: FEMALE RATIO IN GROUP	PHILOPATRY AND BONDING PATTERNS	REPORTED MALE ROLE IN PREDATOR DEFENSE	REFERENCES
Colobus badius	0.54	MP, MB	Multimale defense	21, 22
Procolobus verus	0.63		Not reported	23
Presbytis pileata	0.20	MT, FT	Individual male defense	24
Presbytis entellus	0.17	MT, FB	Lone and multi-male defense	25
Presbytis entellus	0.36	MT, FB	Occasional multi-male defense	26
Cercocebus albigena	0.50	FB	Not reported	27
Cercopithecus aethiops	0.71	MT, FB	Multimale defense	28, 29
Cercopithecus ascanius	0.11	MT, FB	Occasional male defense	30, 31
Miopithecus talapoin	0.48	FB	Male mobbing	32
Erythrocebus patas	0.24	MT, FB	Multimale defense	33, 34
Macaca fasicularis	0.59	MT, FB	Vigilance	35
Macaca mulatta	0.28	MT, FB	Occasional multi-male defense	36
Macaca radiata	0.78	MT, FB	Vigilance, multi-male defense	37
Macaca fuscata	0.33	MT, FB	Occasional multi-male defense	38
Papio hamadryas	0.25	MP	Vigilance, multi-male defense	39, 40
Papio cynocephalus	0.62	MT, FB	Vigilance, multi-male defense	41, 42

TABLE 10.1 *(continued)*

SPECIES	MALE: FEMALE RATIO IN GROUP	PHILOPATRY AND BONDING PATTERNS	REPORTED MALE ROLE IN PREDATOR DEFENSE	REFERENCES
Papio anubis	0.41	MT, FB	Vigilance, multi-male defense	43
Papio ursinus	0.48	MT, FB	Vigilance	44
Pan paniscus	0.50	MP, FB	Not reported	45
Gorilla gorilla	0.33	MT, FT	Individual defense	46
Pan troglodytes	0.37	MP, MB	Cooperative male defense	47, 48

Key:

FB = female bonded
FT = female transfer
MB = male bonded
MP = male philopatric
MT = Male transfer
NMB = nonmale bonded

References:

1. Sussman 1991.
2. Passamani 1995.
3. Digby and Barreto 1993.
4. Digby and Ferrari 1994.
5. Goldizen 1989.
6. Goldizen et al. 1996.
7. Glander 1980.
8. Milton 1982.
9. Crockett and Rudran 1987.
10. Symington 1987.
11. Strier et al. 1993.
12. Janson 1984.
13. Janson 1988.
14. Fedigan 1993.
15. Perry 1996.
16. Boinski 1987a.
17. Boinski 1987b.
18. Mitchell 1994.
19. Boinski 1997.
20. Oates 1977.
21. Struhsaker 1975.
22. Stanford in press a.
23. Oates 1994.
24. Stanford 1991.
25. Newton 1987.
26. Curtin 1975.
27. Waser 1975.
28. Cheney and Seyfarth 1987.
29. Cheney et al. 1988.
30. Struhsaker 1988.
31. Stanford unpublished data.
32. Gautier-Hion 1970.
33. Chism and Rowell 1986.
34. Hall 1965.
35. van Schaik and van Noordwijk 1985.
36. Lindburg 1971.
37. Sugiyama 1971.
38. Maruhashi 1982.
39. Stammbach 1987.
40. Sigg 1980.
41. Altmann et al. 1985.
42. Altmann et al. 1988.
43. Smuts 1985.
44. Bulger and Hamilton 1992.
45. Kano 1992.
46. Harcourt 1978.
47. Goodall 1986.
48. Nishida et al 1990.

vidually. However, primate groups that contain only a few females may simultaneously be multimale, and these males often cooperate to defend the group against predators or invading conspecific males (van Schaik and Hörstermann 1994; Mitani et al. 1996). Male bonding apparently arises in response to a variety of natural selection pressures, including the risk of injury from extragroup conspecifics and injury or death from predators.

While the upper limit on the number of males in a primate group may be set by intrasexual competition among those males, the lower limit appears to be influenced by predation. Anderson (1986) and van Schaik and Hörstermann (1994) found a significant positive correlation between the number of males in a group and the predation risk in the habitat where the group lived.

John Mitani and his colleagues (1996) tested a number of hypotheses about multimale groups. They found that although a species' phylogenetic history influences its mating system, the number of females per group is a major determinant of the number of males. The ratio between the number of males and the number of females per group can itself be a consequence of extrinsic factors such as predation. Figure 10.2 shows the relationship between the adult male-female ratio

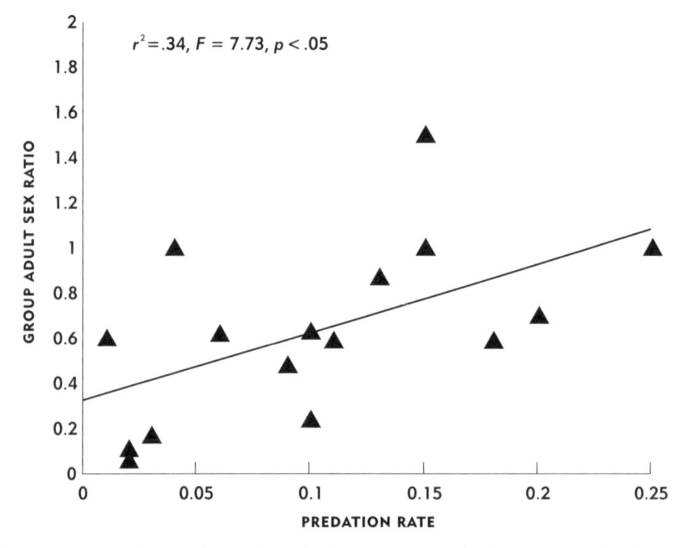

Figure 10.2. Regression of predation rate for 16 primate populations against the adult male-female ratio in the groups. Rate measured as percentage mortality per year due to predation.

and the estimated predation rate (in percentage mortality per year due to predation) for a sample of 16 wild primate populations. There is a significant positive correlation between the number of males per female and the predation rate.

One evolutionary reason for multiple males in primate groups appears to be defense against predators. The presence of more than one male may be advantageous because males are more vigilant than females (Baldellou and Henzi 1992). Although most primatologists have considered only the absolute number of males in the group, that figure is presumably less important than the number of males in relation to the number of group members in need of male protection. The antipredator protection service that males provide may be what enables multimale groups to form around female kin groups in female-philopatric species.

The number or ratio of males may not, however, be as meaningful as what the males do when they encounter a predator. In male-bonded groups, males often repel predators with joint attacks—as do some species in which group males are unrelated (*Cebus* sp., van Schaik and van Noordwijk 1989). Male bonding in Costa Rican squirrel monkeys (*Saimiri oerstedi*) also appears to be a product of communal aggressive male defense against predators (Boinski 1994), though in South American squirrel monkeys (*S. sciureus*) females are philopatric and unrelated males defend against predators (Mitchell 1994). Field evidence supports the idea that predation is an agent of natural selection that promotes male sociality, at least in species most subject to the threat of predation.

Male Sociality and Encounters with Extragroup Males

While male bonding appears to be an adaptive response to predation in some species, the evidence is strong that male bonds are also useful in the acquisition, intragroup control, and intergroup defense of females. Male coalitions form in order to control females to maximize mating opportunities, and also to exclude noncoalition males from mating.

The results of coalitions are twofold. First, large male coalitions potentially are able to dominate small ones during encounters. For this reason natural selection should favor male bonding, including among nonkin, as long as the members of coalitions have greater reproductive success than they would be able to achieve independently. Second, for species that are male bonded because females have a widely dispersed form of sociality, intergroup encounters of rival or neighboring male coalitions occur between groups of both female-bonded and male-bonded species. Males, especially those in highly dimorphic species, use their larger body and canine size to inflict injuries on the males of rival coalitions. Among the cercopithecines, most of which are female philopatric, females tend to be more actively involved than males in intergroup encounters (Cheney 1981), which often involve competition for food. Males are more likely to be involved in intergroup encounters when access to reproduction is at stake (Emlen and Oring 1977; Stanford 1991).

In several primate taxa male coalitions play major roles during intergroup encounters. Chimpanzees are a well-documented case; males of adjacent communities attack one another in coalitions, sometimes with lethal results (Goodall 1986). Because the average degree of kinship among the males of a community is high (Morin et al. 1993), a kin-selected basis exists for such intercommunity male coalitional aggression. Male chimpanzee coalitions, formed for the intragroup control of widely dispersed females, are territorial because of opportunities both to obtain new females and to defend resident females near the territorial border. The evolution of chimpanzee male coalitions may therefore be an arms race in which male cooperation for the purpose of intragroup reproductive control led to increased male bonding. Thereafter, male bonding was enhanced by the need to defend against neighboring male coalitions that are under the same selection pressures.

There is no empirical evidence that male chimpanzees enhance their mating opportunities as a result of participating in such intracommunity male coalitions, but it has been shown that most conceptions occur in the context of temporarily monogamous consortships (Tutin 1979). Thus, access to females is not the only explanation for male chimpanzee coalitions. Male bonobos, also strongly male philopatric, appear to lack the male bonding that is such a prominent part of chimpanzee society. Kano (1992) has observed territorial encounters between male bonobos of neighboring communities that are both aggressive and nonaggressive, but lethal intergroup aggression has not been reported for this species.

The New World analogues of these two African ape social systems are found in the spider monkey and the muriqui. Like chimpanzees, spider monkeys have a fission-fusion polygynous social system in which adult females disperse widely to feed on ripe fruit. In Peruvian black spider monkeys *(Ateles paniscus)*, Symington (1987) found that males form coalitions that both control females and engage in intergroup encounters with male coalitions of other spider monkey groups. Like male chimpanzees, male black spider monkey form dominance hierarchies within the male coalitions. Symington did not, however, observe intergroup male aggression that went beyond chasing.

Among muriquis *(Brachyteles arachnoides)* in Brazil, Strier (1990, 1994) found that males are also related and closely bonded, though their coalitions are not hierarchical and do not appear to function in the control of females. Male muriquis engage in intergroup territoriality, but less aggressively than chimpanzees or spider monkeys. These two neotropical monkeys and the two African pongids are among the largest primates in their geographic areas and therefore are probably less vulnerable to predation than smaller species would be. Male bonding in these taxa appears to be a response to the landscape of female distribution, coupled with the threat posed between groups by males cooperatively monitoring territorial borders.

Among capuchins, communal group defense occurs in a similar pattern, except that the males involved are often unrelated. Male coalitions in various *Cebus* species help to defend females (Robinson 1988, *C. olivaceus;* Perry 1996, *C. capucinus*), are vigilant for marauding and infanticidal extragroup males attempting takeovers (Rose and Fedigan 1995, *C. capucinus*), and work together in predator protection (van Schaik and van Noordwijk 1989, *C. albifrons* and *C. apella*). The exception to this rule of male coalition behavior is *C. albifrons,* in which Janson (1986) found that only the dominant male is active in intergroup encounters. Janson attributed the absence of male cooperation to the monopoly of the dominant male over both food patches and females; the lower-ranking males are left with very little incentive to defend food sources or to help to repel extragroup males.

Once male coalitions have formed to enhance male reproductive success, a second important effect is a potential outcome. This is the mutual defense of male ranges, along with the opportunity to obtain matings with more females than are available in the residents' territory.

Clashes may occur between hostile male coalitions, often at or near the border between two ranges. The larger coalition will be dominant, so that closely bonded coalitions of males become highly beneficial to male reproductive strategies.

Other Male-Bonded Social Mammals

Robert Trivers (1972) hypothesized that male behaviors involved in foraging should not be as strongly selected as behaviors that maximize mating opportunities. Males tend to invest less time and energy in offspring and increase their reproductive success by mating with many females. Females, who invest most of their time and energy in the production and rearing of offspring, benefit most from ensuring the successful survival of those offspring.

Early studies of some of the social carnivores hypothesized that male sociality had evolved primarily to enhance hunting success (Kruuk 1972; Kleiman and Eisenberg 1973). Later evidence showed that other factors were influential in the evolution of male bonds and that groups of carnivores are often larger than would be predicted for optimal hunting success (Kruuk 1972; Packer et al. 1990). Today many social species are thought to form male coalitions in order to defend and to control access to females (Table 10.2). Male bonds appear to exist for this reason among social mammals as diverse as bottle-nosed dolphins (*Tursiops* sp., Connor et al. 1992) and carnivores (Packer et al. 1991).

Both intergroup male aggression and infanticide have been well documented for lions *(Panthera leo)* by Craig Packer, Anne Pusey, and their colleagues. Lions live in fission-fusion social systems that comprise both female and male kin groups, and male coalitions use their cooperative strength to defend their resident pride and to attack others (Packer et al. 1988). Male sociality in this species apparently has not evolved to enhance social hunting success, despite earlier claims (Packer et al. 1990).

Among cheetahs *(Acinonyx jubatus)*, males jointly defend a small portion of the individual ranges of asocial females, attacking male intruders and enlarging their territorial holding through coalitional strength. Caro (1994) showed that male cheetahs do not increase their hunting success through cooperative bonds, nor do coalitions control more territory or obtain access to more females than individual males.

TABLE 10.2. *Features of male sociality among group-living carnivores.*

SPECIES	GROUP MALES RELATED	GROUP FEMALES RELATED	MEAN SIZE OF MALE COALITIONS	MEAN MALE TERRITORY SIZE (SQ. KM)	MALE: FEMALE SEX RATIO	LETHAL INTERGROUP AGRESSION
Panthera leo	Yes	Yes	3.45	203.1	>0.50	Yes
Acinonyx jubatus	Most	—	2.29	37.4	2.29	Yes
Crocuta crocuta	No	Yes	—[a]	60.0	0.78	?
Canis lupus[b]	Yes	Yes	—	>300	>0.50	Yes
Lycaon pictus	Yes	Some	3.66	>1500	1.90	No

Sources: Schaller 1972, Packer et al. 1988 (lion); Caro 1994 (cheetah); Kruuk 1972, Frank 1986 (spotted hyena); Mech 1970 (wolf); Frame et al. 1979 (African wild dog).

a. No coalitions; 16.7 males per group.
b. For Isle Royale.

Coalitions, however, take over territories from residents more effectively, and they pose a greater risk of mortality to other males than lone males do. Herein is one of the main adaptive benefits of male sociality in this species.

Hyenas (*Crocuta crocuta*) engage in aggressive intergroup encounters, in which males participate even though they are normally immigrants and therefore tend to be unrelated. Hans Kruuk's (1972) study of spotted hyenas documented aggressive intergroup encounters in Ngorongoro Crater, from some of which he inferred lethal outcomes. In a later study of spotted hyenas living at much lower density, Frank (1986) did not observe highly aggressive intergroup encounters.

Male bonding is found among canids, in which it plays a role in territorial conflicts. Wolf packs are typically composed of one dominant male-female pair; male coalitions form in the male dominance hierarchy from the offspring of the alpha male (Mech 1970). Encounters between neighboring wolf packs are rare.

Among African wild dogs (*Lycaon pictus*) the social system is similar to that of wolves, with one breeding pair of dominant adults composing the core of the social group. The result is male kin groups and females who typically disperse at maturity (Frame et al. 1979). Male hunting dogs engage in cooperative behavior within the pack, and interpack relations consist of territorial monitoring and chasing of other packs in territorial overlap zones.

Social carnivores thus engage in aggressive intergroup encounters with males of other groups that can lead to severe injury. The attacking males are usually but not necessarily kin, as is the case with primates. The behavior of these coalitional males follows predictions from evolutionary theory of how kin should behave relative to nonkin. Among male Serengeti lions, coalitions are composed mainly, but not entirely, of related males. Although unrelated males cooperate, reproductive success becomes unevenly distributed as coalition sizes increase; relatives forgo mating opportunities to assist each other in larger groups. Nonrelatives apparently do not sacrifice reproductive opportunities for each other (Packer et al. 1991).

The male-bonded social mammals discussed in this chapter all engage in dangerous encounters with extragroup conspecifics or predators or both. Two other male-bonded primates—*Ateles* and *Brachyteles*—who do not engage in highly aggressive territoriality nevertheless utilize male coalitions to defend territories. Not all these species are composed of male kin groups; the multimale polygynous *Cebus* form unrelated male coalitions to repel predators and to guard against male conspecifics.

It remains to be shown for most species that when more males cooperate to repel predators, the hunting success of the predators is indeed lower. For most primate species, antipredator data will be difficult to obtain unless predation events can be observed readily. The question of how to distinguish between male sociality that evolves for predator defense and male bonds that serve mainly to protect against conspecifics remains unanswered. If the behavior patterns resemble one another, and serve similar functions, distinguishing their causes may be impossible. As an example, chimpanzees are extremely combative when engaging in territorial aggression with the members of adjacent communities and may inflict severe injury on their neighbors (Goodall 1986). Many of the same behavioral patterns are seen during hunts of colobus monkeys. Predatory aggression and intergroup conspecific aggression thus bear at least a superficial resemblance to each other and are often distinguished by the context in which they occur.

The defensive behavior of many small monkey species when confronted by a territorial intruder may be different from defense directed at an attacking eagle or big cat, but this remains unknown. Curio (1978) and other ethologists have analyzed avian mobbing by examining its component parts; the same can be done for primates.

It must be borne in mind that defensive behavior, whether directed at predators or at conspecifics, can take a variety of forms. Grave risks are not the entire story, but rather one key component. In spite of the importance of the grave risks that wild primates and other mammals encounter in the wild, the evidence from a wide range of field studies still points to food as playing an important role in ordering primate social systems and molding behavioral strategies.

A final source of mortality needs to be further considered. Infanticide occurs in a number of primate taxa, as well as in other social mammals, and in some populations it accounts for a high percentage of infant deaths. Many observed cases of nonhuman infanticide fit a sexually selected model of the behavior that occurs when extragroup males invade, ousting a resident male and then killing some of the infants in the group who had been fathered by the resident (Hrdy et al. 1995). If male bonds are so essential that they evolved for protection from extragroup threats, why is it that infanticidal species do not live primarily in male-bonded groups, or that Hanuman langur coalitions do not outlast the takeover of one-male groups? Having ousted the resident male, the new residents might remain bonded to repel future male takeovers while sharing reproductive investment among themselves.

It appears that long-lasting male coalitions may not be able to form because Hanuman langur females have opted for a foraging strategy that allows them to be cohesive, as do a range of female-philopatric species. Moreover, the low reproductive value of infants may discourage females from too actively defending them. When predators and extragroup males pose a grave risk, any age class may be taken, though infants may be chosen disproportionately by some predators owing to their ease of capture. For infanticide, however, adults must risk injury to attempt to protect a small amount of reproductive investment. Such a strategy may not be favored, given the risk of injury or death to the defending adult (Clark 1994).

11. Conclusion

THE RELATIONSHIP between chimpanzees and red colobus has serious implications for our understanding of primate social systems. You may question whether the principles of predator-prey ecology apply widely in the primate order. If they do not, then this is a fascinating but anomalous study that has few lessons to teach about the evolutionary role of predation. I have presented evidence of predation's importance among nonhuman primates. I have also pointed out some logical flaws and untested assumptions implicit in models of primate behavior that fail to take into account predation as a driving force. The factors promoting sociality exist in a complicated web of influences that favor different behavioral responses in different circumstances. Further, all species carry evolutionary baggage—their phylogenetic history of adaptations to prior settings—that may constrain their current adaptations to local environments. In this final chapter I return to the questions addressed earlier in the book, to outline the general patterns uncovered during the study as well as the unresolved issues.

1. *To what extent does chimpanzee predation influence the size of the red colobus population, the size of individual red colobus groups, and their age and sex composition? What is the ecological role of chimpanzees as predators?*

Without chimpanzees, the red colobus population at Gombe would be larger and would contain a higher percentage of immatures. In this sense the chimpanzees have a strong influence on the colobus population. Because chimpanzee hunting rates fluctuate widely from year to year and from decade to decade, the impact of hunting on the colobus also varies greatly. Colobus group size is dramatically influenced by predation, as evidenced by the nearly 50 percent size difference between colobus groups in areas subjected to frequent predation versus areas where hunting is rare. All the same, predation by chimpanzees has little effect on the red colobus adult sex ratio. I found little support for the hypothesis that chimpanzees serve as keystone predators in African forests, though more data from a larger sample of sites is needed before we can conclusively state that chimpanzees do not regulate the structure of the ecosystems in which they reside.

2. *What are the most important proximate influences on chimpanzee hunting behavior?*

Chimpanzees at Gombe rarely set out with a prior intention of hunting; prey are attacked when fortuitously encountered during foraging trips for plant foods. The proximate influences on hunting behavior thus include ecological factors that bring both colobus and chimpanzee together and create hunting opportunities, such as the seasonal presence of food trees shared by the two species. The encounter rate is also influenced by the distance the chimpanzees travel each day and therefore is related to their diet. A major nutritional factor is the return in meat that each hunter can expect to obtain, which in turn is based on social factors such as the size and composition of the hunting party. Size and composition also determine the availability of social partners with whom meat can be profitably shared and bartered. Additionally, the presence of swollen females in the party increases the tendency to hunt, regardless of party size.

3. *What do red colobus do to reduce the risk of attack by chimpanzees?*

For arboreal monkeys, it is clearly advantageous to live in large groups if they can thereby avoid or repel predators—even if the group size predisposes the group to higher attack rates. For red colobus, the risk of being attacked by chimpanzees increases sharply in larger groups, but is balanced by the greater protection the colobus receive by virtue of more group members. In habitats where food is not sharply limited, a Gombe red colobus ideally prefers large over small groups. Protection is conferred both by dilution of risk in a large group and by active antipredator defense. Gombe red colobus are not particularly vigilant for predators, and they do not respond to the presence of a potential attacker by moving away rapidly or on subsequent days by avoiding the area in which the predator was encountered.

Colobus apparently gain no advantage by forming mixed-species groups with other arboreal forest monkeys, at least at Gombe. Gombe red colobus alarm call to the sound of chimpanzee pant-hoots more instantaneously than any of the primate species with which they associate. Other species—red-tailed and blue monkeys—flee the scene long before the chimpanzees arrive. Red colobus thus warn other primates of the approach of predators. The red colobus seem to neither benefit by nor suffer from the presence of associated species. This is in sharp contrast to red colobus in West African forests, who capitalize on the presence of guenons as early-warning systems. Both red-tailed and blue monkeys drift in and out of association with colobus throughout the day, gaining the benefit of colobus alarm calls whenever the colobus are feeding in trees compatible with the red-tailed or blue monkey diet. Guenons may also benefit from the chimpanzee preference for catching red colobus, as they are less likely to be targets in the event of an ambush. The red colobus tolerate the presence of other species mainly because they cannot effectively drive them away.

When they are attacked by chimpanzees, the number of male red colobus is an important determinant of their rate of successful defense—up to a point. If many chimpanzees (four or more) are attacking, however, the number of male colobus is not a significant factor in increasing the rate of successful defense. Male red colobus appear to have a kin-selected incentive to defend the females and offspring of the group cooperatively, since without their defense—in those hunts in which no effective counterattack is mounted—the success rate of the chimpanzees is nearly 100 percent.

4. *Why do chimpanzees hunt? Is hunting behavior primarily motivated by nutritional or social factors?*

The nutritional and social influences on hunting are interwoven, though social explanations explain the observed pattern of hunting at Gombe more robustly. The nutrient value of meat and associated body tissues is what chimpanzees seek when they hunt, but their use of meat is not confined to nutrition. Captured carcasses are not simply an end, but a means to an end. Meat is used as a currency in political and reproductive behavior as well. It appears that the uses of meat and the motivation for hunting differ among the several chimpanzee study sites, suggesting that cultural variation exists not only in the pattern of hunting, but also in the causes of hunting behavior.

5. *Has predation by chimpanzees influenced the structure of the red colobus social system?*

The evidence is suggestive that the red colobus social system is an adaptive response to high levels of predation. Defense of the group by multiple males who cooperatively counterattack appears to have been favored by predation. Red colobus are broadly sympatric with chimpanzees across Africa and are the only male-philopatric multimale colobine species. In forests where red colobus occur but chimpanzees do not, the number of male colobus per group is lower. When colobus are attacked by small parties of chimpanzees, the participation of more males increases the rate of successful defense. Thus, Busse's (1978) claim that red colobus social organization is an adaptation to chimpanzee predation may have merit.

The social system is of course the sum of many individual decisions that represent compromise solutions to complex problems. Whether we acknowledge it or not, we are all reductionist in our approach to the natural world, unconsciously categorizing elements of what we see around us. We commonly use the same approach to understanding animal behavior. It works only if we blind ourselves to the multiple levels of analysis that exist, and to the eternal problem of time and scale in evolutionary discussions. Consider this example. Every day, a monkey must decide whether to position himself (or herself) in a central location or in a more peripheral place relative to the rest of his group. By staying near the center of the group, the monkey reduces his food intake by 10 percent from what he could eat at the periphery, owing to the many competitors for food hungrily gobbling the fruits around him. However, by being centrally located the monkey is 20 percent less vulnerable to attack from a predator, because of the additional vigilant eyes scanning for danger. So the monkey should forage in the center of the group, thereby reducing at least one fitness-reducing factor by 10 percent. Or should he? As observers, we need to know much more to be able to predict, or even understand the function of, the individual's spatial position. The influence of one of these two behavior options outweighs the other, meaning that the importance of feeding is not absent but masked by the countervailing benefits of avoiding predation.

There are also many constraints on how the monkey can allocate his activities. Then there is the problem of scale: over how long a period must we watch this monkey and his allies to understand the survival and reproductive values of his behavioral choice? The answer is many generations longer than the duration of most primate field studies. We do not know and cannot at present estimate over what period our monkey benefits from his choices of spatial position. If he dies from predation early in life because of a predilection for feeding peripherally, then to whatever extent his sociality has an evolved basis, he does not pass his slight lack of sociality to his offspring. If he dies later in life, then his genes have already been passed and his death has few if any evolutionary consequences.

Choosing to feed peripherally in order to eat well, on the other hand, is problematic. We do not know if the slight increase in nutrients the monkey thereby obtains makes any difference in his lifetime reproductive output. The natural selection pressure for a particular behavior will also vary over time, according to changes in local ecology, long-term cycles of climate and food availability, and the abundance and behavior patterns of local predators. We are still quite naive about the causes and consequences of the behavior of wild primates, the more so because we tend to focus on the aspects most easily studied. With increasingly long field studies, the roles of predation and other influences that occur too infrequently to be studied in the short term (such as infanticide) should become much clearer.

It is implausible to assume that natural selection exerts an equally strong force at all times and in all places. Very likely, behavior as well as morphology is subject to intense selection pressure during crunch periods that occur infrequently—perhaps only a few years in each century—but that exert a powerful influence. At other times the intensity of natural selection may be much less, as shown in a study of Galápagos Islands finches (Grant and Grant 1989). Even the most ardent adaptationist acknowledges the truth of this pattern. Field researchers typically ignore it because it is untestable, incomprehensible, and potentially confounding in understanding long-term trends. The patterns and overall frequency of predation vary widely over just a few years in the life of the Gombe red colobus population. Logically, the importance of predation as an agent of natural selection must also vary over long periods.

Understanding the role of predation in primate social systems has meaningful conservation implications as well. Just as studies of endangered species seek to establish the natural dietary needs of a population, the way in which predators influence the population structure and social system of these species may be underestimated. Since many endangered species live in habitats where the natural predators no longer exist, their population structure, group size, or other aspects of the social system that may influence successful reproduction may have been unwittingly altered. Understanding the role of predation means understanding the predators, who have complex ecologies and whose dietary preferences sometimes play a major role in structuring the ecological community of other animals.

In this regard it is essential that we gain a better understanding of the relative importance of chronic mortality factors such as food acquisition, and the grave risk of predation. Predation may not influence all species as profoundly as it does red colobus, but the infrequency of its observation belies a long history as an evolutionary influence on ecological communities. As we attempt to fathom social evolution in nonhuman primates, the influence of predation—long a chimera—is slowly becoming clear.

Appendixes: Additional Data on Predator-Prey Ecology

APPENDIX 1. *Studies of predation ecology at Gombe relating to red colobus/chimpanzee, conducted as part of Jane Goodall's long-term research.*

RESEARCHER	TIME	RESEARCH TOPIC
G. Teleki	1968–1970	Hunting behavior of Kasakela chimpanzees
T. Clutton-Brock	1969–1970	Red colobus feeding and ranging ecology
R. Wrangham	1972–1973	Chimpanzee behavioral ecology
C. Busse	1973–1974	Chimpanzee hunting behavior
C. Stanford	1991–1995	Red colobus/chimpanzee predator-prey ecology
S. Watt	1993–1994	Red colobus social behavior and ecology, Busindi Valley
S. Kamenya	1994–1996	Red colobus ecology, Busindi/Mitumba valleys

Appendix 2. *Nonhuman primate species in Gombe National Park, with known population data.*

SPECIES	TAXONOMIC NAME	INDIVIDUALS PER SQ. KM	SOURCE
Chimpanzee	*Pan troglodytes schweinfurthii*	2.5	Goodall 1986; this study
Red colobus	*Colobus badius tephrosceles*	42	This study
Olive baboon	*Papio anubis* spp.	35–40	Gombe Stream Research Center, unpublished data
Red-tailed monkey[a]	*Cercopithecus ascanius schmidti*	34	Stanford, unpublished data
Blue monkey[a]	*Cercopithecus mitis mitis*	8	Stanford, unpublished data
Vervet	*Cercopithecus aethiops*	? (found near human habitations on lakeshore)	
Greater fat-tailed galago	*Galago crassicaudatus*	?	Seen once in 1991 in Mitumba Valley
Needle-clawed galago	*Euoticus elegantulus*	?	Reported by Wrangham in 1975

a. Many hybrids of these two species occur at Gombe.

APPENDIX 3. *Major chimpanzee field studies, with ecological information.*

SITE	HABITAT TYPE	COMMUNITY SIZE	MEAN PARTY SIZE	ESTIMATED RANGE (SQ. KM)	SOURCE
Gombe I, Tanzania (Kasakela)	Riverine forest/woodland/miombo	45	4–6	15–18	Goodall 1986; this study
Gombe II, Tanzania (Mitumba)	Riverine forest/woodland/miombo	25[a]	?	10	This study
Mahale, Tanzania	Riverine forest/woodland	100	6	20–25	Nishida 1990
Taï, Ivory Coast	Lowland rain forest	70[b]	10	27	Cited in Chapman et al. 1994
Kibale, Uganda (Kanyawara)	Middle montane wet forest	50?	5–7	?	Wrangham et al. 1992; Chapman et al. 1994
Mount Assirik, Senegal	Savanna/gallery forest	28	4	>250	Tutin et al. 1983

a. After epidemic in early 1995 that killed at least eight animals.
b. Before recent losses to disease and poaching.

APPENDIX 4. *Hours of observation on each of the red colobus groups.*

GROUP	1991	1992	1993	1994	1995	TOTAL
J	410.25	557.5	202.0	146.5	110.0	1426.25
W	216.5	398.0	115.25	0	0	729.75

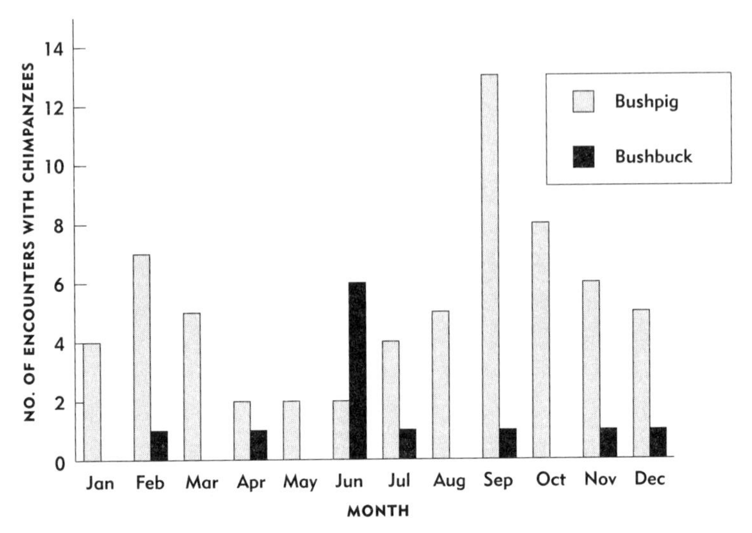

APPENDIX 5. *Monthly variation in the frequency of encounters of chimpanzees with other prey species at Gombe, 1990–1994.*

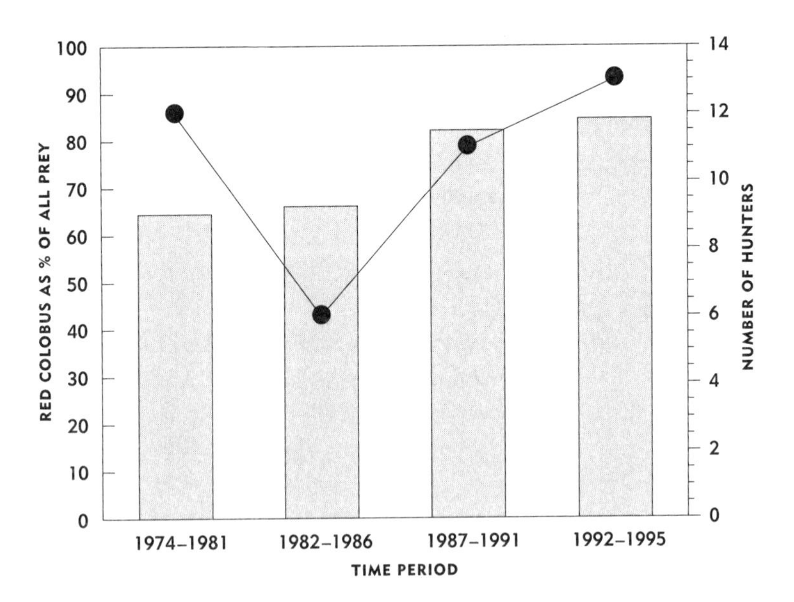

APPENDIX 6. *Changes in the percentages of mammalian prey that were red colobus in relation to the number of adult and adolescent male chimpanzees in the Kasakela community. From Goodall 1986 and Stanford et al. 1994 a.*

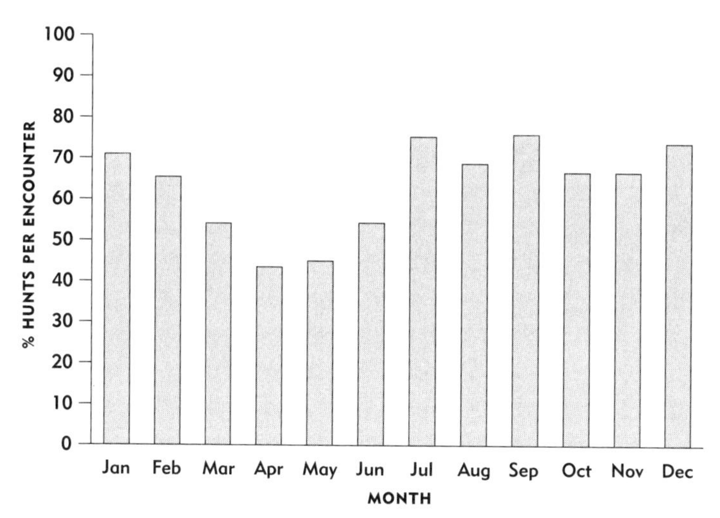

APPENDIX 7. *Monthly variation in the percentage of encounters leading to hunts at Gombe, 1990–1994. Kolmogorov-Smirnov test, $df = 11$, $p = .66$.*

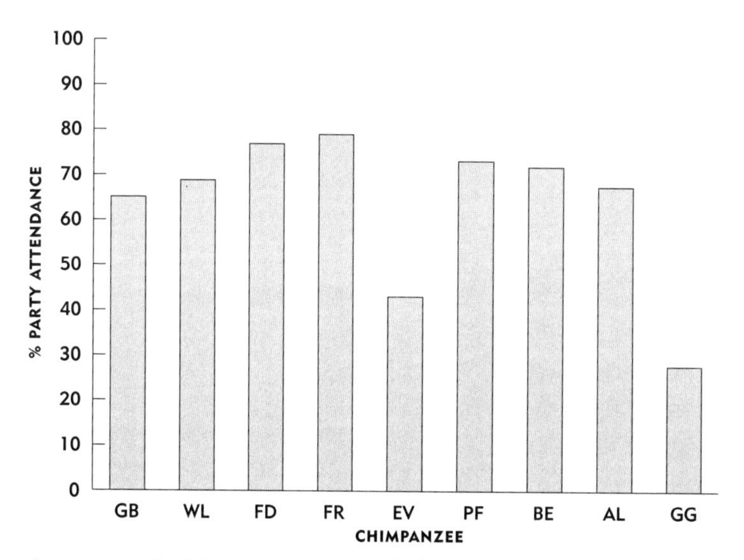

APPENDIX 8. *Variation in sociability among adult hunters at Gombe, measured as the percentage of foraging parties in which each individual's presence was recorded, 1990–1994.*

APPENDIX 9. *Hunting binges at Gombe, 1990–1995. A binge is defined as more than three hunts in a seven-day period.*

DATES	DURATION (DAYS)	MEAN PARTY SIZE	MEAN NO. MALES	NUMBER OF HUNTS	NUMBER OF KILLS
6/13–6/25/90	13	20.8	9.8	5	4
7/7–9/19/90	74	22.1	11.0	38	76
11/19–11/24/90	6	25.6	11.8	5	6
1/17–1/24/91	8	3.8	2.6	5	2
3/14–3/21/91	8	18.5	9.0	4	9
8/12–8/21/91	10	3.9	2.2	10	3
8/31–9/6/91	7	19.8	11.3	6	5
9/12–10/4/91	23	15.6	8.9	14	11
10/16–10/21/91	6	9.3	5.0	4	3
12/8–12/20/91	13	17.3	9.0	7	5
1/15–1/24/92	10	23.3	10.6	7	8
2/14–2/27/92	14	22.5	11.8	8	17
4/18–4/27/92	10	3.0	1.8	4	4
8/24–9/3/92	11	12.2	5.7	9	10
11/6–11/12/92	7	22.3	9.5	4	7
2/8–2/15/93	8	20.5	8.5	4	0
5/2–5/6/93	5	14.3	7.5	4	4
6/16–6/28/93	13	25.6	9.3	7	1
2/11–2/23/94	13	24.7	7.9	9	5
3/5–3/7/94	3	22.0	8.0	4	8
5/12–5/18/94	7	25.0	8.0	4	5
8/24–9/4/94	12	7.6	3.9	8	6
7/11–7/16/95	6	2.8	2.5	4	1

APPENDIX 10. *Age/class distribution of red colobus prey for each hunter at Gombe (more than three observed kills).*

COLOBUS AGE/SEX CLASS	TOTAL	EV	GB	FD	FR	WL	PF	AL	BE	GG	TB	AO	FF
Neonate	44	0	2	4	27	0	3	1	1	1	2	1	0
Infant II	27	1	4	2	5	3	5	0	3	1	2	0	0
Infant of unknown age	57	5	9	8	6	3	5	2	2	3	3	3	2
Juvenile	78	6	12	8	13	10	8	5	6	1	2	2	0
Adult female	19	1	5	1	5	0	1	0	1	0	0	0	1
Adult male	5	0	1	2	1	0	1	0	0	0	0	0	0
Adult of unknown sex	11	3	1	1	4	1	0	0	1	0	0	0	0
Unidentified	55	2	2	5	8	5	2	1	2	0	1	0	1
TOTAL	296	18	26	31	69	22	25	9	16	6	10	6	4

APPENDIX II. *Previous long-term studies of red colobus, with the mean age and sex compositions of groups (range in parentheses), and dispersal pattern.*

RESEARCHER AND AREA STUDIED	TOTAL GROUP SIZE	ADULT MALES	ADULT FEMALES	IMMATURE/ UNIDEN-TIFIED	DISPERSING SEX
Clutton-Brock 1972 (Gombe, Tanzania)	82 (35–110)	11	24	47	?
Struhsaker 1975 (Kibale, Uganda)	34 (9–68)	3–10	8–21	—	Females
Gatinot 1977 (Fathala, Senegal)	25.2 (14–62)	5.5 (4–13)	10.7 (5–27)	9	?
Marsh 1979 (Tana River, Kenya)	18 (12–30)	1.5 (1–2)	9.6 (5–18)	6.9	Males and females
Galat and Galat-Luong 1985 (Taï, Ivory Coast)	32	3	13	16	?
Starin 1991 (Abuko, Gambia)	26	2.5 (2–3)	10.5 (8–13)	13	Females and some males
Davies, cited in Oates 1994 (Tiwai, Sierra Leone)	33	7	13	13	?
This study (Gombe, Tanzania)	27 (8–69)	6.0 (4–14)	11.2 (4–20+)		Females (some males?)

APPENDIX 12. *Age and sex of the five red colobus groups at Gombe, December 1991. See the text for definitions of age categories.*

GROUP	TOTAL GROUP SIZE	ADULT MALES	ADULT FEMALES	SUB-ADULTS	JUVENILES	INFANT II	INFANT I
J	24	5	11	1	3	2	2
W	13	4	5	0	1	2	1
C	20	6	10	0	1	1	2
MK	27	7	14	1	2	1	2
AK	31	8	16	1	2	2	2

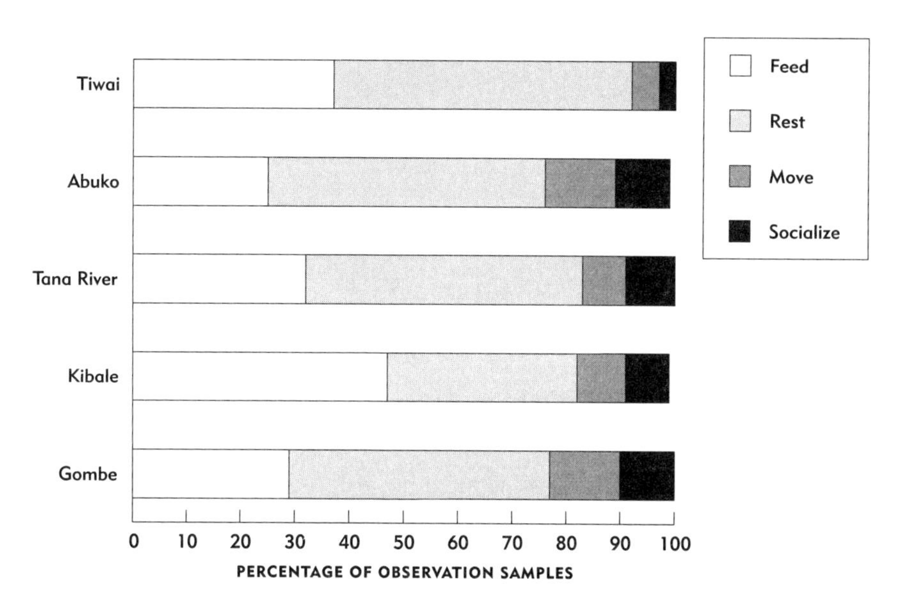

APPENDIX 13. *Activity patterns of Gombe red colobus compared with those of other red colobus populations. Adapted from Oates 1994.*

Appendix 14. *Important plant foods of Gombe red colobus, 1991–1995, based on Clutton-Brock (1975), this study, and Watt (1998). Total feeding scans = 12,950, including 1,813 (14.0%) unidentified.*

SPECIES	FAMILY	LOCAL NAME	% FEEDING SCANS (RANK)	MATURE LEAVES	NEW LEAVES	FRUIT	SEEDS	FLOWERS
Albizzia sp.	Mimosaceae	Msebei	27.8 (1)	x	x		x	
Newtonia buchanani	Mimosaceae	Mka	27.1 (2)	x	x		x	
Sapium ellipticum	Euphorbiaceae	Msasa	11.9 (3)	x	x			
Combretum molle	Combretaceae	Mrama	8.6 (4)	x	x	x	x	x
Pterocarpus angolensis	Papilionaceae	Mninga	8.3 (5)	x	x	x	x	x
Ficus vallis-choudae	Moraceae	Mtobogoro	3.1 (6)	x	x	x	x	
Pseudospondias microcarpa	Anacardiaceae	Mguiza	3.1 (7)	x	x			
Zanha golungensis	Sapindaceae	—	2.0 (8)	x	x			
Pycnanthus angolensis	Myristicaceae	Msulula	1.7 (9)	x	x			
Syzigium guineense	Myrtaceae	Mgege	1.4 (10)	x	x	x	x	
Ficus sp.	Moraceae	Kiholo	1.4 (11)	x	x			
Parinari curatellifolia	Rosaceae	Mbula	1.1 (12)	x	x			
Maesopsis eminii	Rhamnaceae	Mshehe	0.9 (13)	x	x	x	x	
Commiphora madagascariensis	Burseraceae	Mtawera	0.7 (14)	x	x			
Anisophyllea boehmii	Rhizophoraceae	Mshindwe	0.5 (15)	x	x			
Chlorophora excelsa	Moraceae	Mvule	0.2 (16)		x			
Anthocleista schweinfurthii	Loganiaceae	Mrungambale	0.2 (17)	x	x			
Sterculia sp.	Sterculiaceae	—	0.1 (18)	x	x			

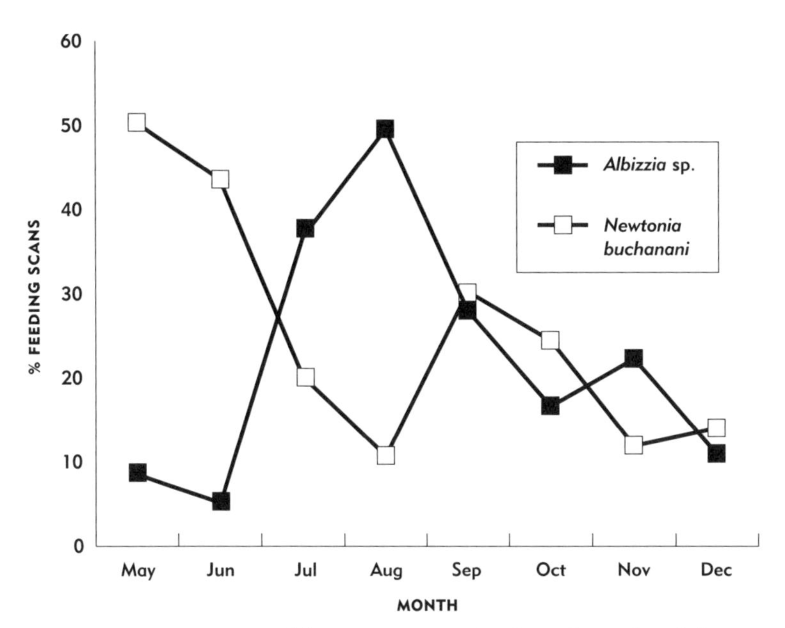

APPENDIX 15. *Monthly variation in feeding by red colobus at Gombe, on* Albizzia *sp. and* Newtonia buchanani, *1991–1995.*

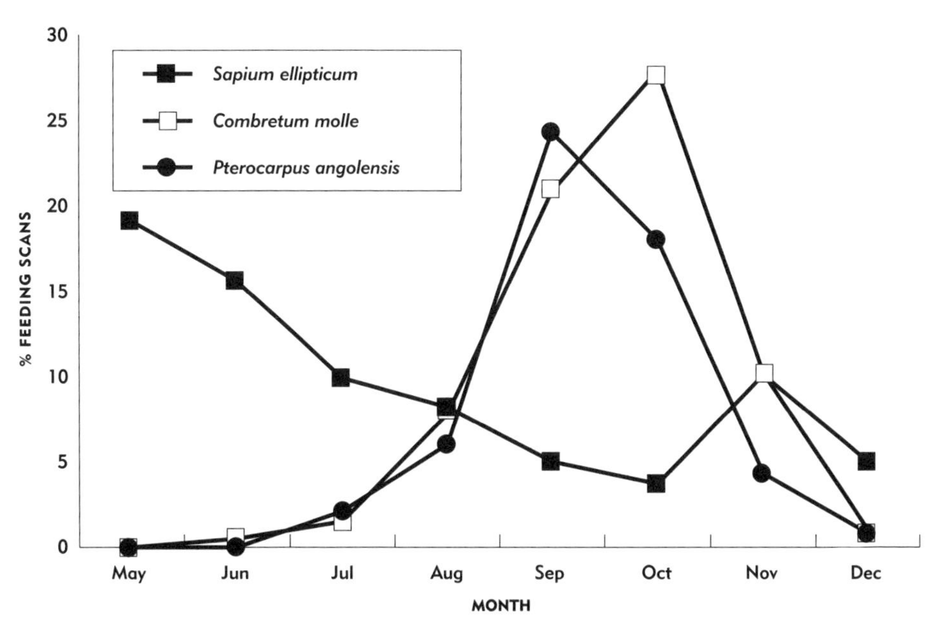

APPENDIX 16. *Monthly variation in feeding by red colobus at Gombe on* Sapium ellipticum, Combretum molle, *and* Pterocarpus angolensis, *1991–1995.*

APPENDIX 17. *Effect of number of colobus counterattackers versus number of chimpanzee attackers on outcome of hunts at Gombe.* N = 48 *hunts in which the number of counterattacking male colobus could be identified.*

NO. MALE CHIMPANZEES ATTACKING	NO. MALE COLOBUS DEFENDING	% SUCCESS OF CHIMPANZEES	SIGNIFICANCE LEVEL
1	< 4	33.3	
1	≥ 4	0.0	.05
2	< 4	25.0	
2	≥ 4	0.0	.01
3	< 4	50.0	
3	≥ 4	25.0	n.s.
4	< 4	83.3	
4	≥ 4	14.3	.01
5	< 4	—	
5	≥ 4	100.0	n.s.
6	< 4	100.0	
6	_ 4	50.0	n.s.
7	< 4	—	
7	≥ 4	33.3	n.s.
8	< 4	75.0	
8	≥ 4	60.0	n.s.
9	< 4	100.0	
9	_ 4	100.0	n.s.
10	< 4	66.7	
10	≥ 4	85.7	.01
11	< 4	100.0	
11	≥ 4	80.0	n.s.
12	< 4	—	
12	≥ 4	83.3	n.s.

Appendix 18. *Demographic changes in red colobus J group, 1991–1995. Offspring of females are listed sequentially below the mother.*

COLOBUS	DEC. 1991	NOV. 1992	AUG. 1993	OCT. 1994	AUG. 1995
JS	AM	AM	AM	AM	AM
JK	AM	AM	AM	AM	AM
JU	AM	AM	Disappeared		
JT	AM	AM	AM	AM	AM
JZ	AM	AM	AM	AM	AM
JB	SAM	AM	AM	AM	AM
JN	AF	AF (b-J6)	AF	AF (b-J9)	AF (b-J10)
J6		INF1-F (b.4/92)	JUVF	Disappeared	
J9				INF1-M (b.6/94) (Predation 9/94)	
J10					INF1-M (b.6/95)
JP	AF (b-JF)	AF	AF (b-J7)	AF	AF
JF	INF2-F	JUVF	JUVF	SAF	AF (b-J14)
J7			INF2-F (b.6/93)	Disappeared	
J14					INF1-M (b.7/95)
JG	AF (b-JH)	AF (b-TJ)	AF	AF (b-J8)	AF
JH	INF2-F	JUVF	Disappeared		
TJ		INF1-M (b.5/92)	Disappeared		
J8				INF1-F (b.9/94)	INF2-F
JR	AF	Predation 10/92			
JJ	INF1-M	Predation 10/92			
JE	AF	AF	AF (b-J11)	AF (b-J13)	AF
JD	JUVM	JUVM	SAM	AM	AM
J11			INF1-M	Disappeared	
J13				INF1-M (b.8/94)	

APPENDIX 18 (*continued*)

COLOBUS	DEC. 1991	NOV. 1992	AUG. 1993	OCT. 1994	AUG. 1995
JA	AF	AF (b-J12)	AF	AF	AF (b-J15)
JY	JUVF	Predation 10/92			
J12		INF1-F (b.10/92)	INF2-F	JUVF	SAF
J15					INF1-M
JM	AF	AF	AF (b-JX)	AF	AF
JO	JUVF	Predation 10/92			
JX			INF1-F	JUVF	JUVF
JV	AF	AF (b-JV) Predation 7/92			
—		INF1 (b.7/92, sex unknown) Predation 7/92			
JQ	AF (b-JI)	AF Predation 10/92			
JI	INF1-M	INF2-M Predation 10/92			
JC	AF	AF (b-J5)	AF	AF (b-JL)	AF
J5		INF2-F (b.6/92) Predation 10/92			
JL				INF2-M	JUVM
JW	SAF	AF (b-J4)	AF	AF	AF (b-J16)
J4		INF1-M (7/92) Predation 10/92			
J16					INF2-F
J2			AF immigrated	AF	AF (b-J17)
J17					INF2-M
J3				AF immigrated	
Group size:	24	20	21	22	26

Key:

AF = adult female
AM = adult male
INF1 = neonate
INF2 = infant
JUVF = juvenile female
JUVM = juvenile male
SAF = subadult female
SAM = subadult male

Appendix 19. *Demographic changes in red colobus W group, 1991–1995. Offspring are listed sequentially below each mother.*

COLOBUS	DEC. 1991	NOV. 1992	AUG. 1993	OCT. 1994	AUG. 1995
WB	AM	AM	AM	With NK group	?
WN	AM	AM	AM	With NK group	?
WK	AM	AM	AM	Disappeared	
WR	AM	AM	AM	Disappeared	
WC	AF (b-WE)	Disappeared			
WE	INF1-F	Disappeared			
WP	AF (b-WG)	AF	AF	Disappeared	
WG	INF2-M	JUV-M	Disappeared		
WL	AF (b-WT)	AF	AF	Disappeared	
WT	INF2-M	JUV-M	Disappeared		
WM	SAF	AF	AF	Disappeared	
WA	AF	AF	AF	Disappeared	
WW	JUV-M	Predation 7/92			
Group size:	13	10	8	2	

Key:
Abbreviations as in Appendix 18.

Appendix 20. *Summary of chimpanzee kills of red colobus in Gombe by year and by valley; data for 1993 are incomplete.*

VALLEY	1990	1991	1992	1993	1994	TOTAL	% TOTAL
Busindi	6	6	0	0	0	12	4.5
Rutanga	5	12	2	6	4	29	10.9
Linda	27	1	19	2	5	54	20.2
Kaskela	0	3	8	0	0	11	4.1
Kakombe	16	9	30	3	18	76	28.5
Mkenke complex	20	10	27	4	4	65	24.3
Kahama complex	8	6	3	0	3	20	7.5
Total	82	47	89	15	34	267	—
% of total	30.7	17.6	33.3	5.6	12.7	—	—

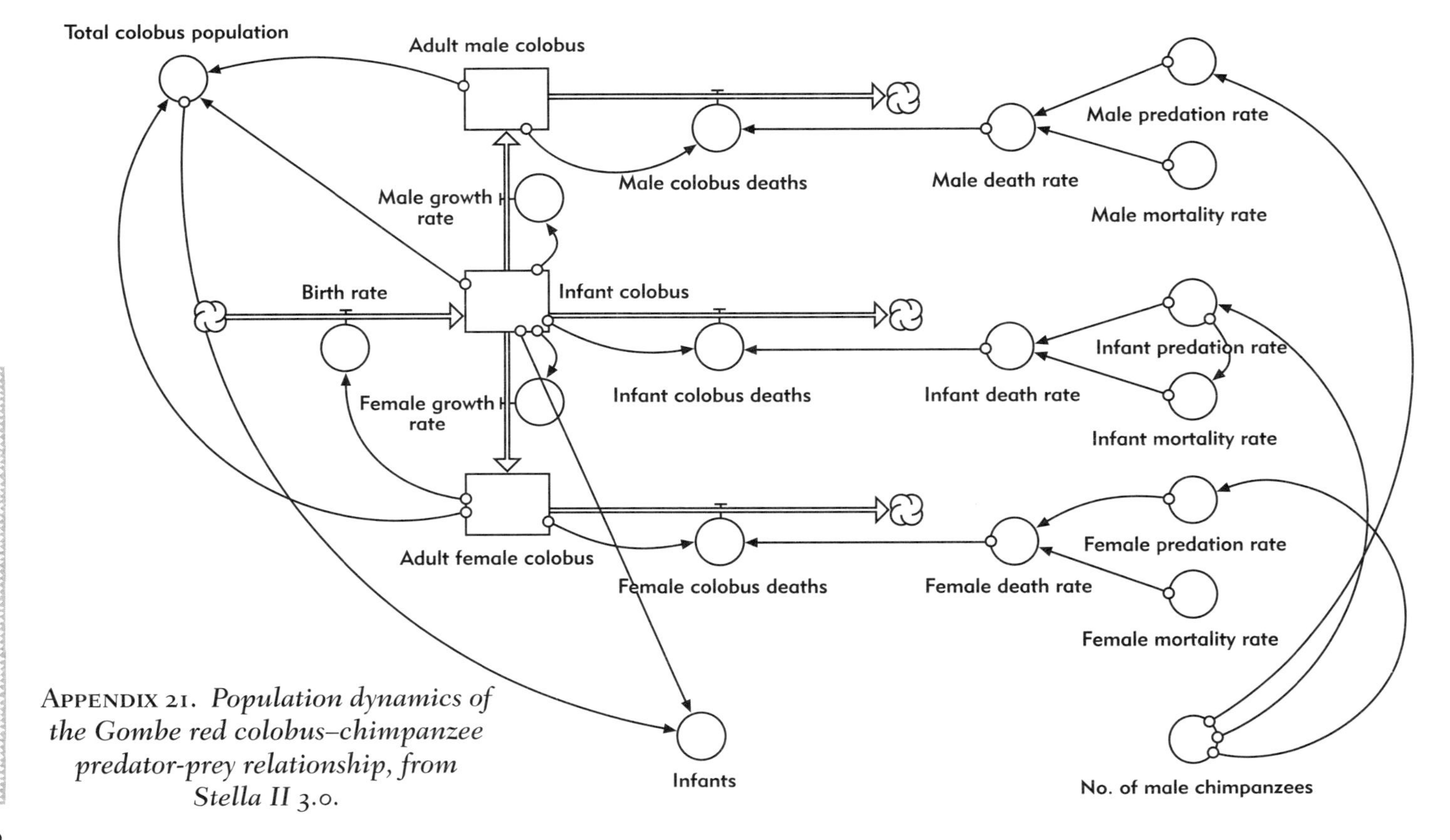

APPENDIX 21. *Population dynamics of the Gombe red colobus–chimpanzee predator-prey relationship, from Stella II 3.0.*

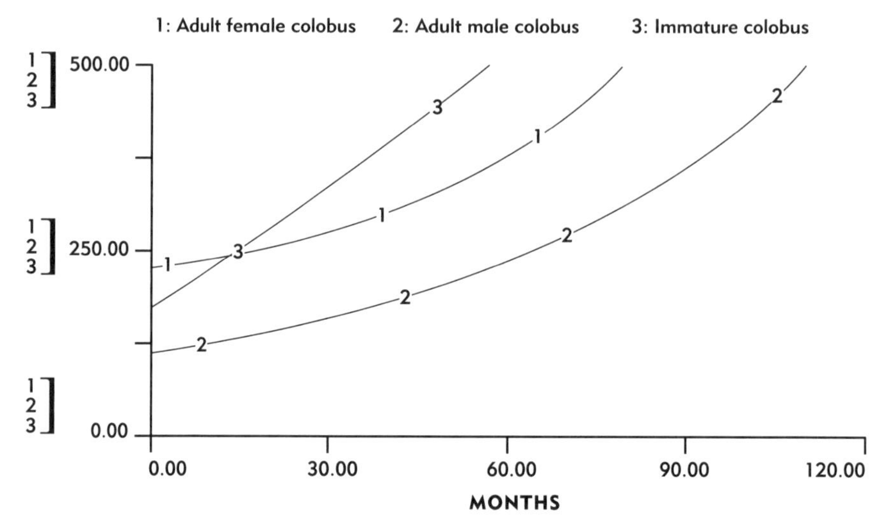

APPENDIX 22. *Computer projection of the increase in the red colobus population over the next ten years if male chimpanzees at Gombe went extinct today.*

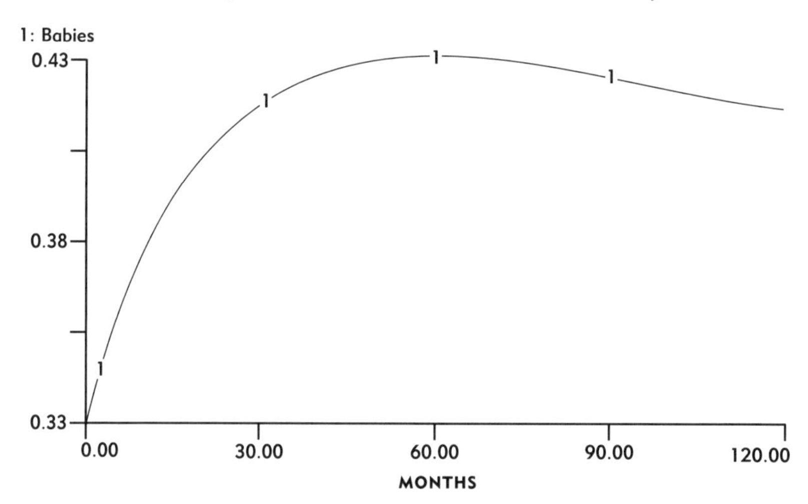

APPENDIX 23. *Computer projection of the increase in the percentage of immatures in the red colobus population over the next ten years if male chimpanzees at Gombe went extinct today.*

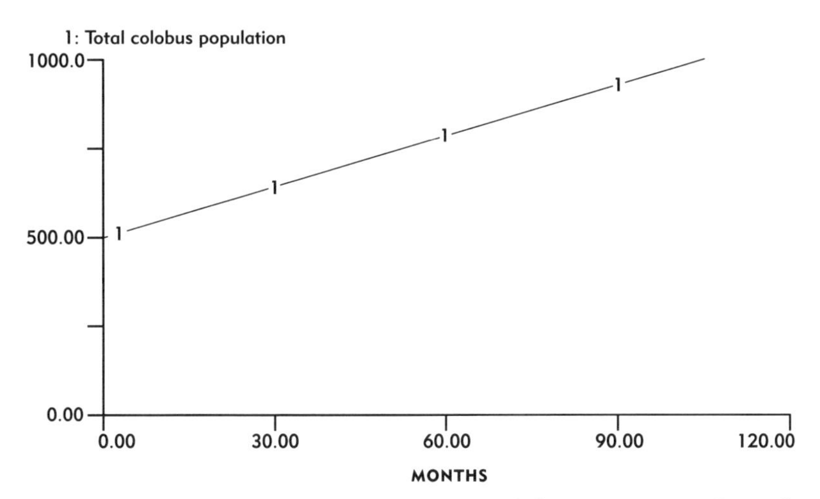

Appendix 24. *Computer projection of the increase in the red colobus population over the next ten years if the number of male chimpanzees at Gombe today were five.*

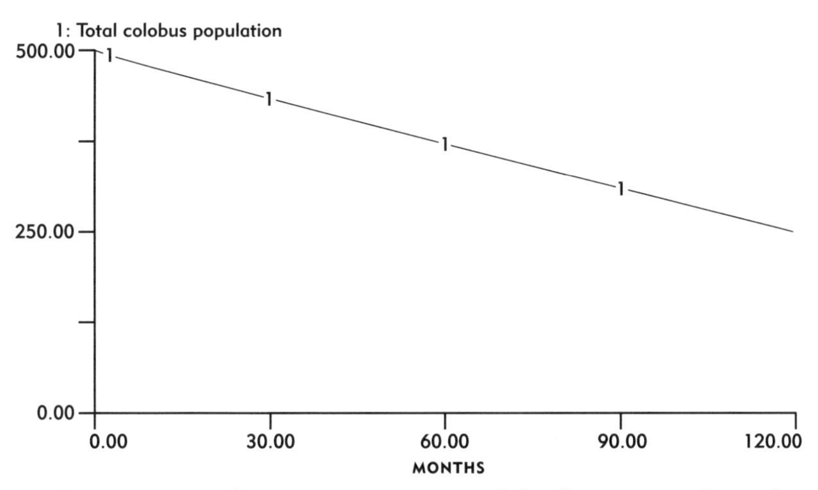

Appendix 25. *Computer projection of the decrease in the red colobus population over the next ten years if the number of male chimpanzees at Gombe today were fifteen.*

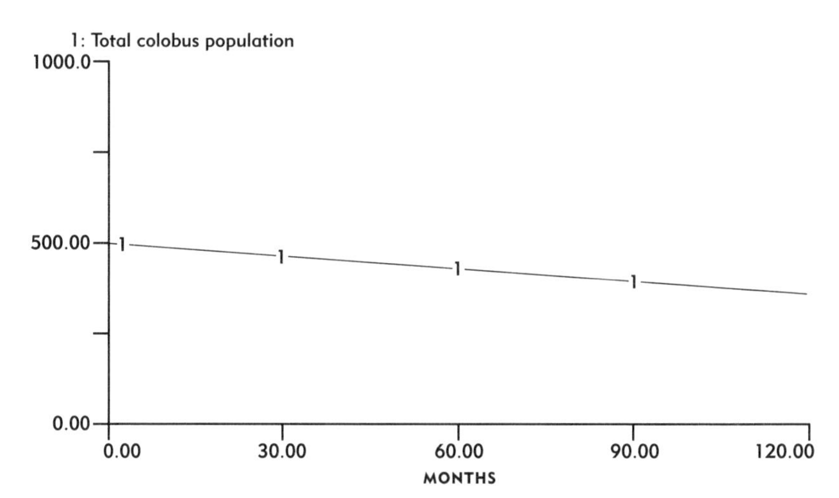

APPENDIX 26. *Computer projection of the decrease in the red colobus population over the next ten years if the number of male chimpanzees at Gombe today were eight and the colobus birthrate were 0.40.*

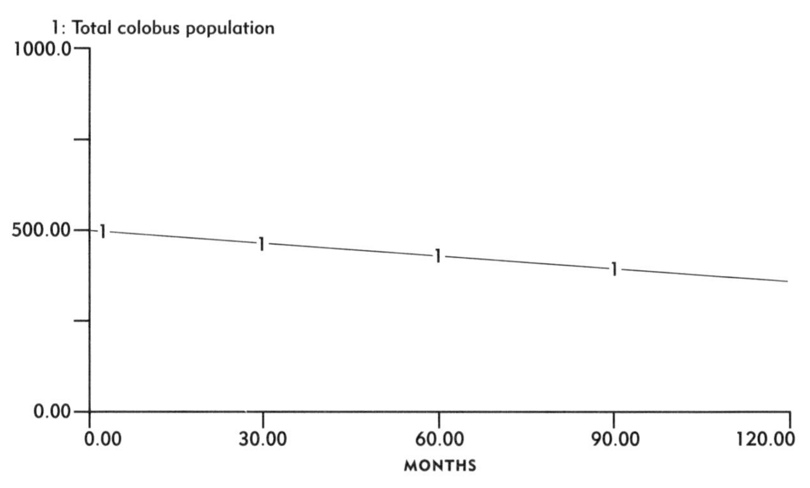

APPENDIX 27. *Computer projection of the growth of the red colobus population over the next ten years if the number of male chimpanzees at Gombe today were eight and the colobus birthrate were 0.54.*

Appendix 28. *Rates of harvesting of oil palm nut* (Elaeis guineensis) *by male and female chimpanzees at Gombe, August–September 1994 and July–August 1995.*

CHIMPANZEE	SEX	TOTAL MINUTES	NUMBER OF PALM NUTS	RATE (NUTS PER MINUTE)
Wilkie	M	168	72	0.43
Beethoven	M	70	42	.60
Frodo	M	96	52	.54
Freud	M	54	12	.22
Atlas	M	49	32	.65
Prof	M	48	25	.52
Male total		485	235	.48
Fifi	F	140	82	.59
Gremlin	F	95	60	.63
Patti	F	51	28	.55
Fanni	F	50	32	.64
Female total		336	202	.60

Appendix 29. *Colobus meat available after hunts at Gombe. N = 60 hunts, successful and unsuccessful, 1991–1995.*

NUMBER OF MALES IN PARTY (ADULT AND ADOLESCENT)	KILOGRAMS OF COLOBUS PER HUNT	KILOGRAMS OF COLOBUS AVAILABLE PER MALE	KILOGRAMS OF COLOBUS AVAILABLE PER INDIVIDUAL	DURATION OF HUNT (MINUTES)	MEAN NUMBER OF KILLS
1	.36	.36	.19	9.7	.4
2	.25	.13	.13	9.0	.3
3	.00	.00	.00	30.5	.0
4	.13	.03	.02	16.9	.3
5	10.00	2.00	.99	23.5	3.5
6	.70	.12	.06	22.7	.6
7	.72	.10	.07	17.7	.4
8	1.53	.19	.09	20.2	.3
9	5.50	.61	.22	29.4	2.2
10	4.75	.48	.20	22.8	1.6
11	1.63	.15	.06	21.8	1.4
12	8.75	.43	.29	29.2	2.5

References

Abrams, P. A. 1994. "Should prey overestimate the risk of predation?" *American Naturalist,* 144: 317–328.

Altmann, J. 1980. *Baboon Mothers and Infants.* Cambridge, Mass.: Harvard University Press.

Altmann, J., G. Hausfater, and S. Altmann. 1985. "Demography of Amboseli baboons." *American Journal of Primatology,* 8: 113–125.

—— 1988. "Determinants of reproductive success in savanna baboons, *Papio cynocephalus.*" In *Reproductive Success* (ed. T. H. Clutton-Brock), pp. 403–418. Chicago: University of Chicago Press.

Altmann, S., and J. Altmann. 1970. *Baboon Ecology.* Chicago: University of Chicago Press.

Alvard, M., and H. Kaplan. 1991. "Procurement technology and prey mortality among indigenous neotropical hunters." In *Human Predators and Prey Mortality* (ed. M. Stiner), pp. 79–104. Boulder, Colo.: Westview Press.

Anadu, P. A., P. O. Elamah, and J. F. Oates. 1988. "The bushmeat trade in southwestern Nigeria: a case study." *Human Ecology,* 16: 199–208.

Anderson, C. M. 1986. "Predation and primate evolution." *Primates,* 27: 15–39.

Axelrod, R., and W. D. Hamilton. 1981. "The evolution of cooperation." *Science,* 211: 1390–96.

Bailey, R. C. and N. R. Peacock. 1988. "Efe pygmies of northeast Zaïre: subsistence strategies in the Ituri Forest." In *Coping with Uncertainty in Food Supply* (ed. I. de Garine and M. A. Harrison), pp. 88–117. Oxford: Clarendon Press.

Baldellou, M. B., and P. Henzi. 1992. "Vigilance, predator detection and the presence of supernumerary males in vervet monkey troops." *Animal Behaviour,* 43: 451–461.

Barnard, C.J. 1981. "Factors affecting flock size mean and variance in a winter population of house sparrows." *Behaviour,* 74: 114–127.

Bartecki, U., and E. W. Heymann. 1987. "Field observation of snake-mobbing in a group of saddle-back tamarins, *Saguinus fuscicollis nigrifrons*." *Folia Primatologica*, 48: 199–202.

Bartlett, T. Q., R. W. Sussman, and J. M. Cheverud. 1993. "Infant killing in primates: a review of observed cases with specific reference to the sexual selection hypothesis." *American Anthropologist*, 95: 958–990.

Bearder, S. 1987. "Lorises, bushbabies, and tarsiers: diverse societies in solitary foragers." In *Primate Societies* (ed. B. B. Smuts, D. L. Cheney, R. M. Seyfarth, R. W. Wrangham, and T. T. Struhsaker), pp. 11–24. Chicago: University of Chicago Press.

Bednarz, J.C. 1988. "Cooperative hunting in Harris' hawks." *Science*, 239: 1525–27.

Bertram, B.C.R. 1980. "Vigilance and group size in ostriches." *Animal Behaviour*, 28: 278–286.

Boesch, C. 1991. "The effects of leopard predation on grouping patterns in forest chimpanzees." *Behaviour*, 117: 220–241.

—— 1994a. "Chimpanzees-red colobus monkeys: a predator-prey system." *Animal Behaviour*, 47: 1135–48.

—— 1994b. "Cooperative hunting in wild chimpanzees." *Animal Behaviour*, 48: 653–667.

—— 1994c. "Hunting strategies of Gombe and Taï chimpanzees." In *Chimpanzee Cultures* (ed. W. C. McGrew, F. B. M. de Waal, R. W. Wrangham, and P. Heltne), pp. 77–92. Cambridge, Mass.: Harvard University Press.

Boesch, C., and H. Boesch. 1989. "Hunting behavior of wild chimpanzees in the Taï National Park." *American Journal of Physical Anthropology*, 78: 547–573.

Boinski, S. 1987a. "Birth synchrony in squirrel monkeys (*Saimiri oerstedi*): a strategy to reduce neonatal predation." *Behavioral Ecology and Sociobiology*, 21: 383–400.

—— 1987b. "Mating patterns in squirrel monkeys (*Saimiri oerstedi*)." *Behavioral Ecology and Sociobiology*, 21: 13–21.

—— 1989. "Why don't *Saimiri oerstedii* and *Cebus capucinus* form mixed-species groups?" *International Journal of Primatology*, 10: 103–114.

—— 1994. "Affiliation patterns among male Costa Rican squirrel monkeys." *Behaviour*, 130: 191–209.

—— 1997. "Behavioral and mtDNA divergence in squirrel monkeys (*Saimiri*)." *American Journal of Physical Anthropology, Supplement* 24: 78–79 (abstract).
Boinski, S., and C. A. Chapman. 1995. "Predation on primates: where are we and what next?" *Evolutionary Anthropology,* 4: 1–13.
Boinski, S., and C.L. Mitchell. 1994. "Male residence and association patterns in Costa Rican squirrel monkeys (*Saimiri oerstedi*)." *American Journal of Primatology,* 34: 157–169.
Brewer, S. 1978. *The Forest Dwellers.* London: Collins.
Brown, L., and D. Amadon. 1968. *Eagles, Hawks and Falcons of the World.* Middlesex, U.K.: Hamlyn Publishing Group.
Brown, L., E. K. Urban, and K. Newman. 1982. *The Birds of Africa,* vol. 1. London: Academic Press.
Bulger, J. and W. Hamilton. 1987. "Rank and density correlates of inclusive fitness measures in a chacma baboon (*Papio ursinus*) troop." *International Journal of Primatology,* 8: 635–650.
Busse, C. D. 1977. "Chimpanzee predation as a possible factor in the evolution of red colobus monkey social organization." *Evolution,* 31: 907–911.
—— 1978. "Do chimpanzees hunt cooperatively?" *American Naturalist,* 112: 767–770.
—— 1980. "Leopard and lion predation upon chacma baboons living in the Moremi Wildlife Reserve." *Botswana Notes Record,* 12: 15–21.
Butynski, T. M. 1982. "Vertebrate predation by primates: a review of hunting patterns and prey." *Journal of Human Evolution,* 11: 421–430.
Cant, J. G. H. 1980. "What limits primates?" *Primates,* 21: 538–544.
Caro, T. M. 1994. *Cheetahs of the Serengeti Plains.* Chicago: University of Chicago Press.
Chapman, C. A. 1986. "Boa constrictor predation and group response in white-faced cebus." *Biotropica,* 18: 171–172.
Chapman, C. A., and L. J. Chapman. 1996. "Mixed-species primate groups in the Kibale Forest: ecological constraints on association." *International Journal of Primatology,* 17: 31–50.
Chapman, C. A., F. J. White, and R. W. Wrangham. 1994. "Party size in chimpanzees and bonobos." In *Chimpanzee Cultures* (ed. W. C. McGrew, F. B. M. de Waal, R. W. Wrangham, and P. Heltne), pp. 41–57. Cambridge, Mass.: Harvard University Press.

Cheney, D. L. 1981. "Intergroup encounters among free-ranging vervet monkeys." *Folia Primatologica,* 35: 124–146.

—— and Seyfarth, R.M. 1981. "Selective forces affecting the predator alarm call of vervet monkeys." *Behaviour,* 76: 25–61.

—— 1987. "The influence of intergroup competition on the survival and reproduction of female vervet monkeys." *Behavioral Ecology and Sociobiology,* 21: 375–386.

—— 1988. Assessment of meaning and the detection of unreliable signals by vervet monkeys. *Animal Behaviour,* 36: 477–486.

—— 1991. *How Monkeys See the World.* Chicago: University of Chicago Press.

Cheney, D. L., and R. W. Wrangham. 1987. "Predation." In *Primate Societies* (ed. B. B. Smuts, D. L. Cheney, R. M. Seyfarth, R. W. Wrangham, and T. T. Struhsaker), pp. 227–239. Chicago: University of Chicago Press.

Cheney, D. L., R. M. Seyfarth, S. J. Andelman, and P. C. Lee. 1988. "Reproductive success in vervet monkeys." In *Reproductive Success* (ed. T. H. Clutton-Brock), pp. 384–402. Chicago: University of Chicago Press.

Chism, J., and T. Rowell. 1986. "Mating and residence patterns of male patas monkeys." *Ethology* 72: 31–39.

Clark, C. W. 1994. "Antipredator behavior and the asset-protection principle." *Behavioral Ecology,* 5: 159–170.

Clutton-Brock, T. H. 1973. "Feeding levels and feeding sites of red colobus *(Colobus badius tephrosceles)* in the Gombe National Park." *Folia Primatologica,* 19: 368–379.

—— 1974. "Activity patterns of red colobus *(Colobus badius tephrosceles).*" *Folia Primatologica,* 21: 161–187.

—— 1975. "Feeding behavior of red colobus and black and white colobus in East Africa." *Folia Primatologica,* 23: 165–207.

Clutton-Brock, T. H., and J. B. Gillett. 1979. "A survey of forest composition in the Gombe National Park, Tanzania." *African Journal of Ecology,* 17: 131–158.

Clutton-Brock, T. H., and P. H. Harvey. 1977. "Primate ecology and social organisation." *Journal of Zoology (London),* 183: 1–39.

Coelho, A., C. A. Bramblett, L. A. Quick, and S. S. Bramblett. 1976. "Resource availability and population density in primates: a socio-

bioenergetic analysis of the energy budgets of Guatemalan howler and spider monkeys." *Primates*, 17: 63–80.
Collins, D. A. 1984. "Spatial pattern in a troop of yellow baboons (*Papio cynocephalus*) in Tanzania." *Animal Behaviour*, 32: 536–553.
Collins, D. A., and W. C. McGrew. 1988. "Habitats of three groups of chimpanzees (*Pan troglodytes*) in western Tanzania compared." *Journal of Human Evolution*, 17: 553–574.
Condit, V. K., and E. O. Smith. 1994. "Predation on a yellow baboon (*Papio cynocephalus cynocephalus*) by a lioness in the Tana River National Primate Reserve, Kenya." *American Journal of Primatology*, 33: 57–64.
Connor, R. C., R. A. Smolker, and A. F. Richards. 1992. "Two levels of alliance formation among male bottlenose dolphins (*Tursiops* sp.)." *Proceedings of the National Academy of Science*, 89: 987–990.
Cords, M. 1987. "Forest guenons and patas monkeys: male-male competition in one-male groups." In *Primate Societies* (ed. B. B. Smuts, D. L. Cheney, R. M. Seyfarth, R. W. Wrangham, and T. T. Struhsaker), pp. 98–111. Chicago: University of Chicago Press.
—— 1990. "Vigilance and mixed-species association of some East African forest monkeys." *Behavioral Ecology and Sociobiology*, 26: 297–300.
—— 1995. "Predator vigilance costs of allogrooming in wild blue monkeys." *Behaviour*, 132: 559–569.
Cowlishaw, G. In press. "Survival or reproduction? The role of vigilance in a desert baboon population." *Behaviour.*
Crockett, C. 1985. "Population studies of red howler monkeys (*Alouatta seniculus*)." *National Geographic Research*, 1: 264–273.
Crockett, C., and R. Rudran. 1987. "Red howler monkey birth data. I. Seasonal variation." *American Journal of Primatology* 13: 347–368.
Crockett, C., and R. Sekulic. 1984. "Infanticide in red howler monkeys (*Alouatta seniculus*)." In *Infanticide: Comparative and Evolutionary Perspectives* (ed. G. Hausfater and S. B. Hrdy), pp. 173–191. New York: Aldine.
Crook, J. H., and S. J. Gartlan. 1966. "On the evolution of primate societies." *Nature*, 210: 1200–3.
Curio, E. 1978. "The adaptive significance of avian mobbing. I. Teleonomic hypotheses and predictions." *Zeitschrift für Tierpsychologie*, 48: 175–183.

Curtin, R. A. 1975. "The socioecology of the common langur, *Presbytis entellus,* in the Nepal Himalaya." Ph.D. diss., University of California, Berkeley.

Daneel, A. B. C. 1979. "Prey size and hunting methods of the crowned eagle." *Ostrich,* 50: 120–121.

Darwin, C. 1859. *The Origin of Species.* London: John Murray.

—— 1871. *The Descent of Man and Selection in Relation to Sex.* London: John Murray.

Davies, A. G. 1994. "Colobine populations." In *Colobine Monkeys: Their Ecology, Behaviour and Evolution* (ed. A. G. Davies and J. F. Oates), pp. 285–310. Cambridge: Cambridge University Press.

Davies, A. G., and J. F. Oates (eds.) 1994. *Colobine Monkeys: Their Ecology, Behaviour and Evolution.* Cambridge: Cambridge University Press.

de Pelham, A., and F. D. Burton. 1976. "More on predatory behavior in nonhuman primates." *Current Anthropology,* 17: 512–513.

DeVore, I., and S. L. Washburn. 1963. "Baboon ecology and human evolution." In *African Ecology and Human Evolution* (ed. F. C. Howell and F. Bourliere), pp. 335–367. Chicago: Aldine.

Diamond, J. M. 1975. "Assembly of species communities." In *Ecology and Evolution of Communities* (ed. M. L. Cody and J. M. Diamond), pp. 342–444. Cambridge, Mass.: Harvard University Press.

Digby, L., and C. Barreto. 1993. "Social organization in a wild population of *Callithrix jacchus.* I. Group composition and dynamics." *Folia Primatologica* 61: 123–134.

Digby, L., and S. Ferrari. 1994. "Multiple breeding females in free-ranging groups of *Callithrix jacchus.*" *International Journal of Primatology* 15: 398–397.

Dittus, W. 1975. "Population dynamics of the toque monkey, *Macaca sinica.*" In *Socioecology and Psychology of Primates* (ed. R. H. Tuttle), pp. 125–151. The Hague: Mouton.

Dunbar, R. I. M. 1988. *Primate Social Systems.* Ithaca: Cornell University Press.

Elgar, M. A. 1989. "Predator vigilance and group size in mammals and birds: a critical review of the empirical evidence." *Biological Reviews,* 64: 13–33.

Emlen, S. T., and L. W. Oring. 1977. "Ecology, sexual selection, and the evolution of mating systems." *Science,* 197: 215–223.

Emmons, L. H. 1987. "Comparative feeding ecology of felids in a neotropical rain forest." *Behavioral Ecology and Sociobiology,* 20: 271–283.
Fanshawe, J. H., and C. D. Fitzgibbon. 1993. "Factors influencing the hunting success of an African wild dog pack." *Animal Behaviour,* 45:479–490.
Fay, J. M., R. Carroll, J. C. Kerbis Peterhans, and D. Harris. 1995. "Leopard attack and consumption of gorillas in the Central African Republic." *Journal of Human Evolution,* 29: 93–99.
Fedigan, L. 1993. "Sex differences and intersexual relations in adult white-faced capuchins, *Cebus capucinus.*" *International Journal of Primatology,* 14: 853–877.
Fisher, R. A. 1958. *The Genetical Theory of Natural Selection,* 2nd ed. New York: Dover.
Fitzgibbon, C. D. 1990. "Mixed-species grouping in Thompson's and Grant's gazelles: the antipredator benefits." *Animal Behaviour,* 39: 1116–26.
Fossey, D. 1983. *Gorillas in the Mist.* Boston: Houghton Mifflin.
Foster, R. B. 1982. "Famine on Barro Colorado Island." In *The Ecology of a Tropical Forest* (ed. E. G. Leigh, A. S. Rand, and D. M. Windsor), pp. 201–212. Washington, D.C.: Smithsonian Institution Press.
Frame, L. H., J. R. Malcolm, G. W. Frame, and H. van Lawick. 1979. "Social organization of the African wild dog *(Lycaon pictus)* in the Serengeti Plains, Tanzania." *Zeitschrift für Tierpsychologie,* 50: 225–249.
Frank, L. 1986. "Social organization of the spotted hyena *(Crocuta crocuta).* I. Demography." *Animal Behaviour,* 34: 1500–9.
Freeland, W. J. 1976. "Pathogens and the evolution of primate sociality." *Biotropica,* 8: 12–24.
Froelich, J. W., Jr., R. W. Thorington, Jr., and J. S. Otis. 1981. "The demography of howler monkeys *(Alouatta palliata)* on Barro Colorado Island, Panama." *International Journal of Primatology,* 2: 207–236.
Fuentes, A. 1995. "Feeding and ranging in the Mentawai Islands langur, *Presbytis potenziani.*" *International Journal of Primatology,* 17: 525–548.
Furuichi, T., and H. Ihobe. 1994. "Variation in male relationships in bonobos and chimpanzees." *Behaviour,* 130: 211–228.

Gagneux, P., D. Woodruff, and C. Boesch. 1997. "Furtive mating in female chimpanzees." *Nature,* 387: 358–359.
Galat-Luong, A. 1983. "Socio-écologie de trois colobes sympatriques, *Colobus badius, C. polykomos* et *C. verus* du Parc National de Taï." Ph.D. diss., University of Paris.
Gatinot, B. L. 1978. "Characteristics of the diet of West African red colobus." In *Recent Advances in Primatology,* vol. 1 (ed. D. J. Chivers and J. Herbert). London: Academic Press.
Gautier-Hion, A. 1970. "L'organisation sociale d'une bande de talapoins (*Miopithecus talapoin*) dans le nord-est du Gabon." *Folia Primatologica,* 12: 116–141.
Gautier-Hion, A., R. Quris, and J. P. Gautier. 1983. "Monospecific vs. polyspecific life: a comparative study of foraging and antipredatory tactics in a community of *Cercopithecus* monkeys." *Behavioral Ecology and Sociobiology,* 12: 325–335.
Ghiglieri, M. P. 1984. *The Chimpanzees of the Kibale Forest.* New York: Columbia University Press.
Glander, K. 1980. "Reproduction and population growth in free-ranging mantled howling monkeys." *American Journal of Physical Anthropology,* 53: 25–36.
—— 1992. "Dispersal patterns in Costa Rican mantled howling monkeys." *International Journal of Primatology,* 13: 415–436.
Goldizen, A. 1989. "Social relationships in a cooperatively polyandrous group of tamarins (*Saguinus fuscicollis*)." *Behavioral Ecology and Sociobiology,* 24: 79–89.
Goldizen, A., M. Mendelson, M. van Vlaardingen, and J. Terborgh. 1996. "Saddle-back tamarin (*Saguinus fuscicollis*) reproductive strategies: evidence from a thirteen-year study of a marked population." *American Journal of Primatology,* 38: 57–83.
Goodall, J. 1963. "Feeding behaviour of wild chimpanzees: a preliminary report." *Symposia of the Zoological Society of London,* 10:39–48.
—— 1968. "Behaviour of free-living chimpanzees of the Gombe Stream area." *Animal Behaviour Monographs,* 1: 163–311.
—— 1971. *In the Shadow of Man.* London: Collins.
—— 1986. *The Chimpanzees of Gombe: Patterns of Behavior.* Cambridge, Mass.: Harvard University Press.
Goodall, J., A. Bandora, E. Bergmann, C. Busse, H. Matama, E. Mpongo, A.Pierce, and D. Riss. 1979. "Inter-community interac-

tions in the chimpanzee population of the Gombe National Park." In *The Great Apes* (ed. D. A. Hamburg and E. R. McCown), pp. 13–53. Menlo Park, Calif.: Benjamin/Cummings.

Goodman, S. M. 1994. "The enigma of anti-predator behavior in lemurs: evidence of a large extinct eagle on Madagascar." *International Journal of Primatology,* 15: 129–134.

Goodman, S. M., S. O'Connor, and O. Langrand. 1993. "A review of predation on lemurs: implications for the evolution of social behavior in small, nocturnal primates." In *Lemur Social Systems and their Ecological Basis* (ed. P. M. Kappeler and J. U. Ganzhorn), pp. 51–66. New York: Plenum Press.

Grant, R. B., and P. R. Grant. 1989. *Evolutionary Dynamics of a Natural Population.* Princeton: Princeton University Press.

Greene, H. W. 1983. "Boa constrictor." In *Costa Rican Natural History* (ed. D. H. Janzen), pp. 380–382. Chicago: University of Chicago Press.

Hall, K. R. L. 1965. "Behaviour and ecology of the wild patas monkey, *Erythrocebus patas,* in Uganda." *Journal of Zoology, London,* 148: 15–87.

Hamilton, W. D. 1971. "Geometry for the selfish herd." *Journal of Theoretical Biology,* 31: 295–311.

Harcourt, A. H. 1978. "Strategies of emigration and transfer by primates, with particular reference to gorillas." *Zeitschrift für Tierpsychologie,* 48: 401–420.

Harley, D. 1985. "Birth spacing in langur monkeys (*Presbytis entellus*)." *International Journal of Primatology,* 6: 227–242.

Hartley, C. W. S. 1977. *The Oil Palm.* New York: John Wiley.

Harvey, P. H., and T. H. Clutton-Brock. 1985. "Life history variation in primates." *Evolution,* 39: 559–581.

Hasegawa, T., M. Hiraiwa, T. Nishida, and H. Takasaki. 1983. "New evidence on scavenging behavior in wild chimpanzees." *Current Anthropology,* 24: 231–232.

Hauser, M. D., and R. W. Wrangham. 1990. "Recognition of predator and competitor calls in nonhuman primates and birds: a preliminary report." *Ethology,* 86: 116–130.

Hausfater, G. 1975. "Dominance and reproduction in baboons (*Papio cynocephalus*)." In *Contributions to Primatology,* vol. 7. Basel: Karger.

Hawkes, K., K. Hill, and J. O'Connell. 1982. "Why hunters gather: optimal foraging and the Ache of eastern Paraguay." *American Ethnologist,* 9: 379–398.

Heymann, E. W. 1990. "Reaction of wild tamarins, *Saguinus mystax* and *Saguinus fuscicollis,* to avian predators." *International Journal of Primatology,* 11: 327–337.

Hill, K., and K. Hawkes. 1983. "Neotropical hunting among the Ache of eastern Paraguay." In *Adaptive Responses of Native Amazonians* (ed. R. B. Hames and W. T. Vickers), pp. 139–188. New York: Academic Press.

Hladik, C. M. 1975. "Ecology, diet and social patterns in Old and New World primates." In *Socioecology and Psychology of Primates* (ed. R. H. Tuttle), pp. 3–35. The Hague: Mouton.

Hohmann, G., and B. Fruth. 1993. "Field observations on meat sharing among bonobos *(Pan paniscus).*" *Folia Primatologica* 60: 225–229.

Holenweg, A. K., R. Noë, and M. Schnabel. 1996. "Waser's gas model applied to associations between red colobus and diana monkeys in the Taï National Park, Ivory Coast." *Folia Primatologica,* 67: 125–136.

van Hooff, J. A. R. A. M., and C. P. van Schaik. 1992. "Cooperation in competition: the ecology of primate bonds." In *Cooperation in Conflict: Coalitions and Alliances in Animals and Humans* (ed. A. H. Harcourt and F. B. M. de Waal), pp. 356–390. Oxford: Oxford University Press.

—— 1994. "Male bonds: affiliative relationships among nonhuman primate males." *Behaviour,* 130: 309–336.

Hoppe-Dominik, B. 1984. "Etude du spectre des proies de la panthère, *Panthera pardus,* dans le Parc National de Tai en Côte d'Ivoire." *Mammalia,* 48: 477–487.

Hrdy, S. B. 1977. *The Langurs of Abu.* Cambridge, Mass.: Harvard University Press.

Hrdy, S. B., C. H. Janson, and C. van Schaik. 1995. "Infanticide: let's not throw the baby out with the bathwater." *Evolutionary Anthropology,* 3: 151–154.

Hudson, J. 1991. "Nonselective small game hunting strategies: an ethnoarchaeological study of Aka pygmy sites." In *Human Predators and Prey Mortality* (Ed. M. Stiner), pp. 105–120. Boulder, Colo.: Westview Press.

Ihobe, H. 1992. "Observations on the meat-eating behavior of wild bonobos *(Pan paniscus)* at Wamba, Republic of Zaïre." *Primates*, 33: 247–250.

Isbell, L. A. 1990. "Sudden short-term increase in mortality of vervet monkeys *(Cercopithecus aethiops)* due to leopard predation in Amboseli National Park, Kenya." *American Journal of Primatology*, 21: 41–52.

—— 1994. "Predation on primates: ecological patterns and evolutionary consequences." *Evolutionary Anthropology*, 3 (2): 61–71.

Isbell, L. A., and T. P. Young. 1993. "Human presence reduces predation in a free-ranging vervet monkey population in Kenya." *Animal Behaviour*, 45: 1233–35.

Izor, R. J. 1985. "Sloths and other mammalian prey of the harpy eagle." In *The Evolution and Ecology of Armadillos, Sloths, and Vermilinguas* (ed. G. G. Montgomery), pp. 343–346. Washington, D.C.: Smithsonian Institution Press.

Janson, C. H. 1984. "Female choice and mating system of the brown capuchin monkey *Cebus apella* (primates: Cebidae)." *Zeitschrift für Tierpsychologie* 65: 177–200.

—— 1985. "Aggressive competition and individual food consumption in wild brown capuchin monkeys *(Cebus apella)*." *Behavioral Ecology and Sociobiology*, 18: 125–138.

—— 1988. "Food competition in brown capuchin monkeys *(Cebus apella)*: quantitative effects of group size and tree productivity." *Behaviour*, 105: 53–76.

—— 1990. "Ecological consequences of individual spatial choice in foraging groups of brown capuchin monkeys, *Cebus apella*." *Animal Behaviour*, 40: 922–934.

—— 1992. "Evolutionary ecology of primate social structure." In *Evolutionary Ecology and Human Behavior* (ed. E. A. Smith and B. Winterhalder), pp. 95–130. New York: Aldine.

Janson, C. H., and M. L. Goldsmith. 1995. "Predicting group size in primates: foraging costs and predation risks." *Behavioral Ecology*, 6: 326–336.

Janson, C. H., and C. P. van Schaik. 1993. "Ecological risk aversion in juvenile primates: slow and steady wins the race." In *Juvenile Primates* (ed. M. E. Pereira and L. A. Fairbanks), pp. 57–74. Oxford: Oxford University Press.

Jorgensen, J. P., and K. H. Redford. 1993. "Humans and big cats as predators in the neotropics." *Symposium of the Zoological Society of London*, 65: 367–390.

Kano, T. 1992. *The Last Ape*. Stanford: Stanford University Press.

Kawanaka, K. 1982. "Further studies on predation by chimpanzees of the Mahale Mountains." *Primates*, 23: 364–384.

Kay, R. N. B., and A. G. Davies. 1994. "Digestive physiology." In *Colobine Monkeys: their Ecology, Behaviour and Evolution* (ed. A. G. Davies and J. F. Oates), pp. 129–150. Cambridge: Cambridge University Press.

Kingdon, J. 1971. *East African Mammals: An Atlas of Evolution in Africa*, vol. 1. New York: Academic Press.

—— 1989. *Island Africa*. Princeton: Princeton University Press.

Kleiman, D. and J. F. Eisenberg. 1973. "Comparisons of canid and felid social systems from an evolutionary perspective." *Animal Behaviour*, 21: 637–659.

Kortlandt, A. 1972. "New perspectives on ape and human evolution." *Stichting voor Psychobiologie*, Amsterdam.

Kruuk, H. 1972. *The Spotted Hyena*. Chicago: University of Chicago Press.

Kuhn, H. J. 1972. "On the perineal organ of male *Procolobus badius*." *Journal of Human Evolution*, 1: 371–378.

Kummer, H. 1968. *Social Organisation of Hamadryas Baboons*. Basel: Karger.

—— 1971. *Primate Societies*. Chicago: Aldine-Atherton.

Lambrecht, J. 1978. "The relationship between food competition and foraging group size in some larger carnivores: a hypothesis." *Zeitschrift für Tierpsychologie*, 46: 337–343.

Leland, L., and T. T. Struhsaker. 1993. "Teamwork tactics." *Natural History*, 102 (4): 42–48.

Leung, W. W. 1968. *Food Composition Table for Use in Africa*. Rome: Food and Agriculture Organization of the United Nations.

Lima, S. L. 1995. "Back to the basics of anti-predatory vigilance: the group-size effect." *Animal Behaviour*, 49: 11–20.

Lima, S. L., and Dill, L. M. 1990. "Behavioral decisions made under the risk of predation: a review and prospectus." *Canadian Journal of Zoology*, 68: 619–640.

Lindburg, D. 1971. "The rhesus monkey in northern India: an ecological and behavioral study." In *Primate Behavior* (ed. L. Rosenblum), pp. 2–106. New York: Academic Press.
MacArthur, R. H. 1960. "On the relation between reproductive value and optimal predation." *Proceedings of the National Academy of Sciences,* 46: 143–145.
Marler, P. 1970. "Vocalizations of East African monkeys. I. Red colobus." *Folia Primatologica,* 13: 81–91.
Marsh, C. W. 1979. "Female transference and mate choice among Tana River red colobus." *Nature,* 281: 568–569.
—— 1981. "Time budget of Tana River red colobus." *Folia Primatologica,* 35: 30–50.
—— 1986. "A resurvey of Tana primates and their forest habitat." *Primate Conservation,* 7: 72–81.
Maruhashi, T. 1982. "An ecological study of troop fissions of Japanese monkeys *(Macaca fuscata yakui)* on Yaku Island, Japan." *Primates,* 23: 317–337.
Masui, K., Y. Sugiyama, A. Nishimura, and H. Ohsawa. 1975. "The life table of Japanese monkeys at Takasakiyama." In *Contemporary Primatology: Proceedings of the Fifth Congress of the International Primatological Society* (ed. M. Kawai, S. Kondo, and A. Ehara). Basel: Karger.
McGrew, W.C. 1992. *Chimpanzee Material Culture.* Cambridge: Cambridge University Press.
Mech, D. 1970. *The Wolf: The Ecology and Behavior of an Endangered Species.* Minneapolis: University of Minnesota Press.
Melnick, D. J., and M. Pearl. 1987. "Cercopithecines in multimale groups: genetic diversity and population structure." In *Primate Societies* (ed. B. B. Smuts, D. L. Cheney, R. M. Seyfarth, R. W. Wrangham, and T. T. Struhsaker), pp. 121–134. Chicago: University of Chicago Press
Mills, M. G. L., and H. C. Biggs. 1993. "Prey apportionment and related ecological relationships between large carnivores in Kruger National Park." *Symposium of the Zoological Society of London,* 65: 253–268.
Milton, K. 1980. *The Foraging Strategy of Howler Monkeys: A Study in Primate Economics.* New York: Columbia University Press.

—— 1982. "Dietary quality and demographic regulation in a howler monkey population." In *The Ecology of a Tropical Forest* (ed. E. G. Leigh, A. S. Rand, and D. M. Windsor), pp. 273–289. Washington, D.C.: Smithsonian Institution Press.

Mitani, J. C., and K. L. Brandt. 1994. "Social factors influence the acoustic variability in the the long-distance calls of male chimpanzees." *Ethology*, 96: 233–252.

Mitani, J. C., J. Gros-Louis, and J. H. Manson. 1996. "Number of males in primate groups: comparative tests of competing hypotheses." *American Journal of Primatology*, 38: 315–332.

Mitchell, C. L. 1994. "Migration alliances and coalitions among adult male South American squirrel monkeys (*Saimiri sciureus*)." *Behaviour*, 130: 169–190.

Moore, J. 1984. "Female transfer in primates." *International Journal of Primatology*, 5: 537–589.

—— 1996. "Savanna chimpanzees, referential models and the last common ancestor." In *Great Ape Societies* (ed. W. C. McGrew, L. F. Marchant, and T. Nishida), pp. 275–292. Cambridge: Cambridge University Press.

Morbeck, M. E., and A. L. Zihlmann. 1989. "Body size and proportions in chimpanzees, with special reference to *Pan troglodytes schweinfurthii* from Gombe National Park, Tanzania." *Primates*, 30: 369–382.

Mori, A. 1979. "Analysis of population changes by measurement of body weight in the Koshima troop of Japanese monkeys." *Primates*, 20: 371–397.

Morin, P. A., J. Wallis, J. Moore, R. Chakraborty, and D. S. Woodruff. 1993. "Non-invasive sampling and DNA amplification for paternity exclusion, community structure, and phylogeography in wild chimpanzees." *Primates*, 34: 347–356.

Morris, K., and J. Goodall. 1977. "Competition for meat between chimpanzees and baboons of the Gombe National Park." *Folia Primatologica*, 28: 109–121.

Moynihan, M. 1962. "The organization and probable evolution of some mixed species flocks of neotropical birds." *Smithsonian Miscellaneous Collections*, 143: 1–40.

Mturi, F. A. 1991. "The feeding ecology and behaviour of the red colobus monkey (*Colobus badius kirkii*)." Ph.D. diss., University of Dar es Salaam, Tanzania.

Muller, M., E. Mpongo, C. B. Stanford, and C. Boehm. 1995. "A note on the scavenging behavior of wild chimpanzees." *Folia Primatologica,* 65: 43–47.

Munn, C. A. 1985. "Permanent canopy and understory flocks in Amazonia: species composition and population density." In *Neotropical Ornithology* (ed. P. A. Buckley, M. S. Foster, E. S. Morton, R. S. Ridgely, and F. G. Buckley), pp. 683–712. *Ornithological Monographs,* 36.

Munn, C. A., and J. W. Terborgh. 1979. "Multispecies territoriality in neotropical foraging flocks." *Condor,* 81: 338–347. National Research Council. 1981. *Techniques in Primate Population Ecology.* Washington, D.C.: National Academy Press.

Newton, P. N. 1987. "The social organization of forest Hanuman langurs." *International Journal of Primatology,* 8: 199–232.

—— 1988. "The variable social organization of Hanuman langurs *(Presbytis entellus),* infanticide, and the monopolization of females." *International Journal of Primatology,* 9: 59–77.

Nishida, T. 1968. "The social group of wild chimpanzees in the Mahali Mountains." *Primates,* 9: 167–224.

—— 1972. "A note on the ecology of the red colobus monkeys *(Colobus badius tephrosceles)* living in the Mahali Mountains." *Primates,* 13: 57–64.

—— 1990. "A quarter century of research in the Mahale Mountains: an overview." In *The Chimpanzees of the Mahale Mountains* (ed. T. Nishida), pp. 3–36. Tokyo: University of Tokyo Press.

Nishida, T., S. Uehara, and R. Nyondo. 1983. "Predatory behavior among wild chimpanzees of the Mahale Mountains." *Primates,* 20: 1–20.

Nishida, T., M. Hiraiwa-Hasegawa, T. Hasegawa, and Y. Takahata. 1985. "Group extinction and female transfer in wild chimpanzees in the Mahale National Park, Tanzania." *Zeitschrift für Tierpsychologie,* 67: 284–301.

Nishida, T., H. Takasaki, and Y. Takahata. 1990. "Demography and reproductive profiles." In *The Chimpanzees of the Mahale Mountains* (ed. T. Nishida), pp. 63–97. Tokyo: University of Tokyo Press.

Nishida, T., T. Hasegawa , H. Hayaki, Y. Takahata, and S. Uehara. 1992. "Meat-sharing as a coalition strategy by an alpha male chimpanzee." In *Topics in Primatology,* vol. 1 (ed. T. Nishida, W. C. McGrew, P.

Marler, and M. Pickford), pp. 159–174. Tokyo: University of Tokyo Press.

Noë, R. 1992. "Alliance formation among male baboons: shopping for profitable partners." In *Cooperation in Conflict: Coalitions and Alliances in Animals and Humans* (ed. A. H. Harcourt and F. B. M. de Waal), pp. 285–322. Oxford: Oxford University Press.

Noë, R., and R. Bshary. 1997. "The formation of red colobus–diana monkey associations under predation pressure from chimpanzees." *Proceedings of the Royal Society of London,* ser. B, 264: 253–259.

Oates, J. F. 1977. "The social life of a black-and-white colobus monkey." *Zeitschrift für Tierpsychologie,* 45: 1–60.

—— 1994. "The natural history of African colobines." In *Colobine Monkeys: Their Ecology, Behaviour and Evolution* (ed. A. G. Davies and J. F. Oates), pp. 75–128. Cambridge: Cambridge University Press.

—— 1996. "Habitat alteration, hunting and the conservation of forests of folivorous primates in African forests." *Australian Journal of Ecology,* 21: 1–9.

Oates, J. F., and G. H. Whitesides. 1990. "Association between olive colobus (*Procolobus verus*), diana guenons (*Cercopithecus diana*), and other forest monkeys in Sierra Leone." *American Journal of Primatology,* 21: 129–146.

Oates, J. F., G. H. Whitesides, A. G. Davies, P. G. Waterman, S. M. Green, G. L. Da Silva, and S. Mole. 1990. "Determinants of variation in tropical forest primate biomass: new evidence from West Africa." *Ecology,* 71: 328–343.

Overdorff, D. J. 1995. "Life-history and predation in *Eulemur rubriventer* in Madagascar." *American Journal of Physical Anthropology,* suppl. 20: 164 (abstract).

Owings, D. H, and D. F. Hennessy. 1984. "The importance of variation in sciurid visual and vocal communication." In *The Biology of Ground-Dwelling Squirrels* (ed. J. O. Murie and G. R. Michener), pp. 169–200. Lincoln: University of Nebraska Press.

Packer, C. and L. Ruttan. 1988. "The evolution of cooperative hunting." *American Naturalist,* 132: 159–198.

Packer, C., L. Herbst, A. E. Pusey, J. D. Bygott, J. P. Hanby, S. J. Cairns, and M. Borgerhoff-Mulder. 1988. "Reproductive success in lions." In

Reproductive Success (ed. T. H. Clutton-Brock), pp. 363–383. Chicago: University of Chicago Press.

Packer, C., D. Scheel, and A. E. Pusey. 1990. "Why lions form groups: food is not enough." *American Naturalist,* 136: 1–19.

Packer, C., D. A. Gilbert, A. E. Pusey, and S. J. O'Brien. 1991. "A molecular genetic analysis of kinship and cooperation in African lions." *Nature,* 351: 562–565.

Pagel, M. and P. H. Harvey. 1993. "Evolution of the juvenile period in mammals." In *Juvenile Primates* (ed. M. E. Pereira and L. A. Fairbanks), pp. 28–37. Oxford: Oxford University Press.

Passamani, M. 1995. "Field observation of a group of Geoffroy's marmosets mobbing a margay cat." *Folia Primatologica,* 64: 163–166.

Peetz, A., M. A. Norconk, and W. G. Kinzey. 1992. "Predation by jaguar on howler monkeys *(Alouatta seniculus)* in Venezuela." *American Journal of Primatology,* 28: 223–228.

Pereira, M. E., and L. A. Fairbanks. 1993. "What are juvenile primates all about?" In *Juvenile Primates* (ed. M. E. Pereira and L. A. Fairbanks), pp. 3–15. Oxford: Oxford University Press.

Peres, C. A. 1993. "Anti-predation benefits in a mixed-species groups of tamarins." *Folia Primatologica,* 61: 61–76.

Perry, S. 1996. "Intergroup encounters in wild white-faced capuchins *(Cebus capucinus).*" *International Journal of Primatology,* 17: 309–330.

Pitman, C. R. S. 1974. *A Guide to the Snakes of Uganda,* 2nd ed. Codicote, U.K.: Wheldon and Wesley.

Power, M. 1991. *The Egalitarians, Human and Chimpanzee.* Cambridge: Cambridge University Press

Pulliam, H. R. 1973. "On the advantages of flocking." *Journal of Theoretical Biology,* 38: 419–422.

Ransom, T. W. 1972. "Ecology and social behavior of baboons in the Gombe National Park." Ph.D. diss., University of California, Berkeley.

Rettig, N. L. 1978. "Breeding behavior of the harpy eagle *(Harpia harpyia).*" *Auk,* 95: 629–643.

Robbins, M. M. 1995. "A demographic analysis of male life history and social structure of mountain gorillas." *Behaviour* 132: 21–48.

Robinson, J. 1988. "Demography and group structure in wedge-capped capuchin monkeys, *Cebus olivaceous.*" *Behaviour,* 104: 202–231.

Robinson, J., and C. H. Janson 1987. "Capuchins, squirrel monkeys, and atelines: socioecological convergence with Old World primates." In *Primate Societies* (ed. B.B. Smuts, D.L. Cheney, R.M. Seyfarth, R.W. Wrangham, and T.T. Struhsaker), pp. 69–82. Chicago: University of Chicago Press.

Rodman, P. 1973. "Population composition and adaptive organization among orang-utans of the Kutai Reserve." In *Comparative Ecology and Behaviour of Primates* (ed. R.P. Michael and J. H. Crook), pp. 171–209. London: Academic Press.

Ron, T., S. P. Henzi, and U. Motro. 1996. "Do female chacma baboons compete for safe spatial position in a southern woodland habitat?" *Behaviour,* 133: 475–490.

Rose, L. M. 1994. "Benefits and costs of resident males to females in white-faced capuchins, *Cebus capucinus." American Journal of Primatology,* 32: 235–248.

Rose, L. M., and L. M. Fedigan. 1995. "Vigilance in white-faced capuchins, *Cebus capucinus,* in Costa Rica." *Animal Behaviour,* 49: 63–70.

de Ruiter, J. R. 1986. "The influence of group size on predator scanning and foraging behaviour of wedgecapped capuchin monkeys (*Cebus olivaceus*)." *Behaviour,* 98: 240–258.

Ruvolo, M., T. R. Disotell, M. W. Allard, W. M. Brown, and R. L. Honeycutt. 1991. "Resolution of the African hominoid trichotomy by use of a mitochondrial gene sequence." *Proceedings of the National Academy of Science,* 88: 1570–74.

Sabater Pi, J., M. Bermejo, G. Ilera, and J. J. Vea. 1993. "Behavior of bonobos (*Pan paniscus*) following their capture of monkeys in Zaïre." *International Journal of Primatology,* 14: 797–804.

Sade, D. S., K. Cushing, P. Cushing, J. Dunaif, A. Figueroa, J. R. Kaplan, C. Laurer, D. Rhodes, and J. Schneider. 1976. "Population dynamics in relation to social structure on Cayo Santiago." *Yearbook of Physical Anthropology,* 20: 253–262.

van Schaik, C. P. 1983. "Why are diurnal primates living in groups?" *Behaviour,* 87: 120–144.

—— and R. I. M. Dunbar. 1990. "The evolution of monogamy in large primates: a new hypothesis and some crucial tests." *Behaviour,* 115: 30–62.

—— and J. van Hooff. 1983. "On the ultimate causes of primate social systems." *Behaviour,* 85: 91–117.
—— and M. Hörstermann. 1994. "Predation risk and the number of adult males in a primate group: a comparative test." *Behavioral Ecology and Sociobiology,* 35: 261–272.
—— and T. Mitrasetia. 1990. "Changes in the behaviour of wild long-tailed macaques *(Macaca fascicularis)* after encounters with a model python." *Folia Primatologica,* 55: 104–108.
—— and M. A. van Noordwijk. 1985. "Evolutionary effect of the absence of felids on the social organization of the macaques on the island of Simeulue (*Macaca fascicularis,* Miller 1903)." *Folia Primatologica,* 44: 138–47.
—— and M. A. van Noordwijk. 1989. "The special role of male *Cebus* monkeys in predation avoidance and its effect on group composition." *Behavioral Ecology and Sociobiology,* 24: 265–76.
Schaller, G. 1967. *The Deer and the Tiger.* Chicago: University of Chicago Press.
—— 1972. *The Serengeti Lion.* Chicago: University of Chicago Press.
Schleidt, W. M. 1973. "Tonic communication: continual effects of discrete signs in animal communication." *Journal of Theoretical Biology,* 42: 359–386.
Seidensticker, J. 1983. "Predation by *Panthera* cats and measures of human influence in habitats of South Asian monkeys." *International Journal of Primatology,* 4: 323–326.
Seidensticker, J., and C. McDougal. 1993. "Tiger predatory behaviour, ecology and conservation." *Symposia of the Zoological Society of London,* 65: 105–125.
Sherman, P. W. 1977. "Nepotism and the evolution of alarm calls." *Science,* 197: 1246–53.
Sigg, H. 1980. "Differentiation of female positions in hamadryas one-male-units." *Zeitschrift für Tierpsychologie,* 53: 265–302.
Sigg, H., and A. Stolba. 1981. "Home range and daily march in a hamadryas baboon troop." *Folia Primatologica,* 36: 40–75.
Skorupa, J. 1989. "Crowned eagles *Stephanoetus coronatus* in rainforest: observations on breeding chronology and diet at a nest in Uganda." *Ibis,* 131: 294–298.
Smuts, B. B. 1985. *Sex and Friendship in Baboons.* New York: Aldine.

Sommer, V. 1994. "Infanticide among the langurs of Jodhpur: testing the sexual selection hypothesis with a long-term record." In *Infanticide and Parental Care* (ed. S. Parmigiani and F. vom Saal), pp. 155–198. London: Harwood Academic Publishers.

Sommer, V., A. Srivastava, and C. Borries. 1992. "Cycles, sexuality and conception in free-ranging langurs *(Presbytis entellus)*." *American Journal of Primatology*, 28: 1–27.

Stammbach, E. 1987. "Desert, forest, and montane baboons: multilevel societies." In *Primate Societies* (ed. B. B. Smuts, D. L. Cheney, R. M. Seyfarth, R. W. Wrangham, and T. T. Struhsaker), pp. 112–120. Chicago: University of Chicago Press.

Stanford, C.B. 1989. "Predation by jackals *(Canis aureus)* on capped langurs *(Presbytis pileata)* in Bangladesh." *American Journal of Primatology*, 18: 53–56.

—— 1991a. "Behavioral ecology of the capped langur in Bangladesh: reproductive tactics in one-male groups." In *Contributions to Primatology*, vol. 26. Basel: Karger.

—— 1991b. "Social dynamics of intergroup encounters in the capped langur *(Presbytis pileata)*." *American Journal of Primatology*, 25: 35–48.

—— 1994a. "Notes on raptor migration in western Tanzania." *Scopus*, 18: 1–5.

—— 1994b. "A note on reproduction and neonate dispersal in the African rock python *(Python sebae)*." *Herpetological Review*, 25 (3): 125.

—– 1995a. "The influence of chimpanzee predation on group size and anti-predator behaviour in red colobus monkeys." *Animal Behaviour*, 49: 577–587.

—— 1995b. "Chimpanzee hunting behavior and human evolution." *American Scientist*, 83: 256–261.

—— 1996. "The hunting ecology of wild chimpanzees: implications for the behavioral ecology of Pliocene hominids." *American Anthropologist*, 98: 96–113.

—— In press a. "Predation, and male bonds in primate societies." *Behaviour*.

—— In press b. *The Hunting Apes: Meat-Eating and the Origin of Human Behavior*. Princeton: Princeton University Press.

—— In press c. "The social behavior of chimpanzees and bonobos: empirical evidence and shifting assumptions." *Current Anthropology.*

Stanford, C. B., and P. Msuya. 1995. "An annotated list of the birds of Gombe National Park, Tanzania." *Scopus,* 19: 38–46.

Stanford, C. B., J. Wallis, H. Matama, and J. Goodall. 1994a. "Patterns of predation by chimpanzees on red colobus monkeys in Gombe National Park, Tanzania, 1982–1991." *American Journal of Physical Anthropology,* 94: 213–228.

Stanford, C. B., J. Wallis, E. Mpongo, and J. Goodall. 1994b. "Hunting decisions in wild chimpanzees." *Behaviour,* 131: 1–20.

Starin, E. D. 1991. "Socioecology of the red colobus monkey in the Gambia with particular reference to female-male differences and transfer patterns." Ph.D. diss., City University of New York.

Starin, E. D., and G. Burghardt. 1992. "African rock pythons *(Python sebae)* in the Gambia: observations on natural history and interactions with primates." *The Snake* 24: 50–62.

Stephens, D. W., and J. R. Krebs. 1986. *Foraging Theory.* Princeton: Princeton University Press.

Steyn, P. 1982. *Birds of Prey of Southern Africa.* Cape Town: David Phillip.

Strier, K. B. 1990. "New World primates, new frontiers: insights from the woolly spider monkey, or muriqui *(Brachyteles arachnoides).*" *International Journal of Primatology,* 11: 7–19.

—— 1994. "Brotherhoods among atelins: kinship, affiliation, and competition." *Behaviour,* 130: 151–167.

Strier, K. B., F. Mendes, J. Rimoli, and A. Rimoli. 1993. "Demography and social structure of one group of muriquis *(Brachyteles arachnoides).*" *International Journal of Primatology,* 14: 513–526.

Strong, D. R., Jr., D. Simberloff, L. G. Abele, and A. B. Thistle (Eds.). 1984. *Ecological Communities: Conceptual Issues and the Evidence.* Princeton: Princeton University Press.

Struhsaker, T. T. 1975. *The Red Colobus Monkey.* Chicago: University of Chicago Press.

—— 1981. "Vocalizations, phylogeny, and paleogeography of red colobus monkeys *(Colobus badius).*" *African Journal of Ecology,* 19: 265–284.

—— 1988. "Male tenure, multi-male influxes, and reproductive success in redtail monkeys (*Cercopithecus ascanius*)." In *A Primate Radiation: Evolutionary Biology of the African Guenons* (ed. A. Gautier-Hion, F. Bourliere, J. Gautier, and J. Kingdon), pp. 340–363. Cambridge: Cambridge University Press.

Struhsaker, T. T., and M. Leakey. 1990. "Prey selectivity by crowned hawk-eagles on monkeys in the Kibale Forest, Uganda." *Behavioral Ecology and Sociobiology*, 26: 435–443.

Struhsaker, T. T., and L. Leland. 1985. "Infanticide in a patrilineal society of red colobus monkeys." *Zeitschrift für Tierpsychologie*, 9: 89–132.

—— 1987. "Colobines: infanticide by adult males." In *Primate Societies* (ed. B. B. Smuts, D. L. Cheney, R. M. Seyfarth, R. W. Wrangham, and T. T. Struhsaker), pp. 83–97. Chicago: University of Chicago Press.

Struhsaker, T. T., and J. F. Oates. 1975. "Comparison of the behavior and ecology of red colobus and black-and-white colobus monkeys in Uganda: a summary." In *Socioecology and Psychology of Primates* (ed. R. H. Tuttle), pp. 103–123. The Hague: Mouton.

Struhsaker, T. T., and T. R. Pope. 1991. "Mating system and reproductive success: a comparison of two African forest monkeys (*Colobus badius* and *Cercopithecus ascanius*)." *Behaviour*, 117: 182–205.

Strum, S. 1976. "Predatory behavior of olive baboons (*Papio anubis*) at Gilgil, Kenya." Ph.D. diss., University of California, Berkeley.

—— 1982. "Agonistic dominance in male baboons; an alternative view." *International Journal of Primatology*, 3: 175–202.

Sugiyama, Y. 1971. "Characteristics of the social life of bonnet macaques (*Macaca radiata*)." *Primates*, 12: 247–266.

Sullivan, K. A. 1985. "The advantages of social foraging in downy woodpeckers." *Animal Behavior*, 32: 16–22.

Sussman, R. 1991. "Demography and social organization of free-ranging *Lemur catta* in the Beza Mahafaly Reserve, Madagascar." *American Journal of Primatology*, 15: 45–67.

Symington, M. M. 1987. "Ecological and social correlates of party size in the black spider monkey, *Ateles paniscus chamek*." Ph.D. diss., Princeton University.

Takahata, Y., T. Hasegawa, and T. Nishida. 1984. "Chimpanzee predation in the Mahale Mountains from August 1979 to May 1982." *International Journal of Primatology*, 5: 213–233.

Tattersall, I. 1982. *The Primates of Madagascar.* New York: Columbia University Press.
Teleki, G. 1973. *The Predatory Behavior of Wild Chimpanzees.* Lewisburg, Pa.: Bucknell University Press.
—— 1977. "Still more on predatory behavior in nonhuman primates." *Current Anthropology,* 18:107–108.
—— 1981. "The omnivorous diet and eclectic feeding habits of chimpanzees in Gombe National Park, Tanzania." In *Omnivorous Primates: Gathering and Hunting in Human Evolution* (Ed.by R. S. O. Harding and G. Teleki), pp. 303–343. New York: Columbia University Press.
Terborgh, J. 1983. *Five New World Primates.* Princeton: Princeton University Press.
—— 1988. "The big things that run the world—a sequel to E. O. Wilson." *Conservation Biology,* 2: 402–403.
—— 1990. "Mixed flocks and polyspecific associations: costs and benefits of mixed groups to birds and monkeys." *American Journal of Primatology,* 21: 87–100.
Terborgh, J., and C. H. Janson. 1986. "The socioecology of primate groups." *Annual Review of Ecology Systematics,* 17: 111–135.
Tilson, R. L., and R. R. Tenaza. 1977. "Social organization of simakobu monkeys (*Nasalis concolor*) in Siberut Island, Indonesia." *Journal of Mammalogy,* 58: 202–212.
Treves, A. In press. "Has predation shaped the social systems of arboreal primates?" *International Journal of Primatology.*
Trivers, R. T. 1972. "Parental investment and sexual selection." In *Sexual Selection and the Descent of Man, 1871–1971* (ed. B. Campbell), pp. 136–179. Chicago: Aldine.
Tsukahara, T. 1993. "Lions eat chimpanzees: the first evidence of predation by lions on wild chimpanzees." *American Journal of Primatology,* 29: 1–11.
Tutin, C. E. G. 1979. "Mating patterns and reproductive strategies in a community of wild chimpanzees (*Pan troglodytes schweinfurthii*)." *Behavioral Ecology and Sociobiology,* 6: 29–38.
Uehara, S. 1986. "Sex and group differences in feeding on animals by wild chimpanzees in the Mahale Mountains National Park, Tanzania." *Primates,* 27: 1–13.

—— 1997. "Predation on mammals by the chimpanzee *(Pan troglodytes)*: an ecological review." *Primates,* 38: 193–214.

Uehara, S., T. Nishida, M. Hamai, T. Hasegawa, H. Hayaki, M. Huffman, K. Kawanaka, S. Kobayoshi, J. Mitani, Y. Takahata, H. Takasaki, and T. Tsukahara. 1992. "Characteristics of predation by the chimpanzees in the Mahale Mountains National Park, Tanzania." In *Topics in Primatology,* vol. 1, *Human Origins* (ed. T. Nishida, W. C. McGrew, P. Marler, M. Pickford, and F. B. M. de Waal), pp. 143–158. Tokyo: University of Tokyo Press.

Vega-Redondo, F., and O. Hasson. 1993. "A game-theoretic model of predator-prey signalling." *Journal of Theoretical Biology,* 162: 309–319.

Vermeij, G. J. 1982. "Unsuccessful predation and evolution." *American Naturalist,* 120: 701–720.

Wachter, B., M. Schnabel, and R. Noë. In press. "Diet overlap and poly-specific associations of red colobus and diana monkeys in the Taï National Park, Ivory Coast." *Ethology.*

Wallis, J. 1997. "A survey of reproductive parameters in the free-ranging chimpanzees in Gombe National Park." *Journal of Reproduction and Fertility,* 109: 297–307.

Wallis, J., and H. Matama. 1993. "Social and environmental factors influencing sleep/wake patterns of wild chimpanzees." *American Journal of Primatology,* 30: 354 (abstract).

Waser, P. M. 1975. "Monthly variation in feeding and activity patterns of the mangabey, *Cercopithecus albigena* (Lydekker)." *East African Wildlife Journal,* 13: 249–263.

—— 1982. "Primate polyspecific associations: do they occur by chance?" *Animal Behavior,* 30: 1–8.

Waser, P. M., and T. J. Case. 1981. "Monkeys and matrices: on the coexistence of "omnivorous" forest primates." *Oecologia,* 49: 102–108.

Waterman, P. G., and G. M. Choo. 1981. "The effects of some digestibility-reducing compounds in leaves on food selection by some Colobinae." *Malaysian Applied Biology,* 10: 147–162.

Watts, D. P. 1989. "Infanticide in mountain gorillas: new cases and a reconsideration of the evidence." *Ethology,* 81: 1–18.

Whitesides, G. H. 1989. "Interspecific associations of diana monkeys, *Cercopithecus diana,* in Sierra Leone, West Africa: biological significance or chance?" *Animal Behaviour,* 37: 760–776.

Wiens, J. A. 1989. *The Ecology of Bird Communities*, vol. 2, *Processes and Variations*. Cambridge: Cambridge University Press.

Wilson, E. O. 1987. "The little things that run the world (The importance and conservation of invertebrates)." *Conservation Biology*, 1: 344–346.

Wrangham, R. W. 1975. "Behavioural ecology of chimpanzees in Gombe National Park, Tanzania." Ph.D. diss., Cambridge University.

—— 1979. "On the evolution of great ape social systems." *Social Science Information*, 18: 335–368.

—— 1980. "An ecological model of female-bonded primate groups." *Behaviour*, 75: 262–292.

Wrangham, R. W., and D. Peterson. 1996. *Demonic Males*. Boston: Houghton Mifflin.

Wrangham, R. W., and B. B. Smuts. 1980. "Sex differences in the behavioural ecology of chimpanzees in the Gombe National Park, Tanzania." *Journal of Reproduction and Fertility*, suppl. 28: 13–31.

Wrangham, R. W., and E. van Zinnicq Bergmann-Riss. 1990. "Rates of predation on mammals by Gombe chimpanzees, 1972–1975." *Primates*, 31: 157–170.

Wrangham, R. W., N. L. Conklin, G. Etot, J. Obua, K. D. Hunt, M.D. Hauser, and A. P. Clark. 1993. "The value of figs to chimpanzees." *International Journal of Primatology*, 14: 243–256.

Wright, P. C. 1984. "Biparental care in *Aotus trivirgatus* and *Callicebus moloch*." In *Female Primates: Studies by Women Primatologists* (ed. M. F. Small), pp. 59–75. New York: Alan R. Liss.

—— 1985. "The costs and benefits of nocturnality for *Aotus trivirgatus* (the night monkey)." Ph.D. diss., City University of New York.

—— 1995. "Demography and life history of free-ranging *Propithecus diadema edwardsi* in Ranomafana National Park, Madagascar." *American Journal of Physical Anthropology*, suppl. 20: 224 (abstract).

Wright, S. J., M. E. Gompper, and B. DeLeon. 1994. "Are large predators keystone species in neotropical forests? The evidence from Barro Colorado Island." *Oikos*, 71: 279–294.

Yasukawa, K. 1981. "Song repertoire in the red-winged blackbird (*Agelaius phoeniceus*): a test of the Beau Geste hypothesis." *Animal Behavior*, 29: 114–125.

Yerkes, R. M. 1941. “Conjugal contrasts among chimpanzees.” *Journal of Abnormal and Social Psychology,* 36: 175–199.

Yost, J. A., and P. M. Kelley. 1983. “Shotguns, blowguns, and spears: the analysis of technological efficiency.” In *Adaptive Responses of Native Amazonians* (ed. R. B. Hames and W. T. Vickers), pp. 189–224. New York: Academic Press.

Zuberbühler, K., R. Noë, and R. M. Seyfarth. In press. “Diana monkey long-distance calls: messages for conspecifics and predators.” *Animal Behaviour.*

Zuckerman, S. 1932. *The Social Life of Monkeys and Apes.* New York: Harcourt Brace.

Acknowledgments

VERY LITTLE IS KNOWN about the predator-prey dynamics of any wild primate population. A bit more is known about a few primate populations, thanks to a number of field researchers. Many of these individuals helped me with the research and writing of this book, and I needed all of them. While *Chimpanzee and Red Colobus* builds on previous research at Gombe by a number of people, my primary debt is to Dr. Jane Goodall, whose pioneering role and continuing vision have enriched the world view of more people than have all other primatologists combined. She not only discovered meat-eating and hunting by wild chimpanzees in the early 1960s, but has carried Gombe Stream Research Center through its fourth decade as the site where wild animals have been studied longest.

Since the late 1960s several other scientists who studied either chimpanzees or red colobus at Gombe have been interested in predation, and I cite their research findings continually throughout this book: Geza Teleki, Richard Wrangham, Curt Busse, and Timothy Clutton-Brock. The ongoing projects in the Mahale Mountains of Tanzania, led by Toshisada Nishida, and in the Taï forest in Ivory Coast, implemented by Christophe and Hedwige Boesch, have been fascinating counterpoints in comparing hunting styles and effects among distant chimpanzee populations.

During 18 months studying langurs in forests of Bangladesh and India for my doctoral dissertation in the late 1980s, I believed conducting a primate field study to be an utterly solitary venture. I realize now that doing fieldwork can be a more exciting enterprise when one has on-the-spot colleagues and friends with whom to share results and of whom to ask advice. Gombe has never had a shortage of insightful and supportive people, both Tanzanian and expatriate, and to them I am very grateful. I especially thank Anthony Collins and Janette Wallis, who while coordinating research at Gombe pointed me toward some key findings and provided much-needed friendship. I also thank wardens Peter Msuya, Stephan Qolli, and Dattomax Selenyika for their official help and friendship, and John Dota, Leah Gardner, Linda Marchant, William McGrew, James Murray, Alnazir Haji Mohammed, Barbara Smuts, Charlotte Uhlenbroek, and William Wallauer for their companionship and discussion in the field.

The doctoral research projects of Sharon Watt and Shadrack Kamenya on the Mitumba Valley colobus groups are the latest knowledge of Gombe red colobus socioecology, and I appreciate their allowing me to discuss some of their results here. I am grateful for the hospitality of Toshisada Nishida during a trip to Mahale National Park in 1991, and of Richard Wrangham during a brief visit to Kibale National Park in 1995. Also at Kibale, Thomas Struhsaker kindly supplied me with taped vocalizations of other red colobus populations. The Jane Goodall Research Center at the University of Southern California, which archives data and film footage from Gombe, provided access to the filmed and written records of many hunts that occurred when I was not in the park.

The unsung heroes of Gombe are certainly the Tanzanian research assistants, some of whom have observed chimpanzees continuously since the 1960s, and all of whom know the forest and the lives of wild chimpanzees better than virtually anyone else. Foreign researchers come and go, but without the *watafiti,* research at Gombe could not continue. Headed by Hilali Matama, the team has included Eslom Mpongo, Hamisi Mkono, Yahaya Alamasi, Selemani Yahaya, Gabo Paulo, Bruno Herman, the late Msafiri Katoto, Issa Salala, the late David Mussa, Karoli Alberto, Tofficki Mikidaddi, Madua Juma, Nasibu Sadiki, and Methodi Vyampi.

For six years of financial support I thank the L. S. B. Leakey Foundation (1990–1991), the Fulbright Foundation (1992–1993), the National Geographic Society (1993–1994), and the University of Southern California (1993, 1995). I also thank the offices of Tanzania National Parks, the Tanzanian Commission for Science and Technology, and the Serengeti Wildlife Research Institute for permission to conduct the research.

For their helpful critical discussion and their advice on various sections of this book, I owe a debt of gratitude to Christopher Boehm, Guy

Cowlishaw, Michael Huffman, Lynne Isbell, William McGrew, Joseph Manson, Erin Moore, Jim Moore, Martin Muller, Charles Roseman, Karen Strier, and Sharon Watt. I thank Stephen Lansing and Martin Muller for help with demographic modeling. For reading and commenting on the entire manuscript I am grateful to Russell Tuttle and Richard Wrangham. And I thank Michael Fisher of Harvard University Press for his editorial advice.

Since 1983 I have made 14 trips to the tropics to watch primates, for up to 16 months at a stretch. Since 1985 these trips have often taken me away from my wife, Erin, and more recently from our young children. For their support during all of these forced separations, and for the trips we have made together, I am deeply thankful.

Index